安定化原理

赤沼 篤夫

Stabilization Principle

東京図書出版

Contents —— 目次

Stabilization Principle

Prologue

Subjective existences which are conceptual existences in the human mind are stable and are perceived with scholar values. But objective existences are natural and independent from the human conception and have some instability. They are subject to the natural regulations which are essentially contingent. Individual natural events and existences are substantial products from the contingency and are controlled by the regulations induced from the contingency. Essentially events stabilize according to statistical probability. Stabilization means that events change to gain status with the higher statistical probabilities. The local maximum statistical probability brings forth the stability. Position and motion are ought to be handled with not scholar values but statistical values. Physical events with mass have been usually statistically examined but those without mass are ought to be examined. Events which nature brings forth with or without mass should be examined. The nature is actualized by the stabilization which the contingency brings forth. Events follow the stabilization processes increasing the probabilities of existence and persistence. Hence natural events have some instabilities. The statistical perception on the essentialities of natural events developed stabilization principle. This principle consists of the following two axioms of existence and persistence.

1. Insecurity; Events exist. Existence involves some statistical insecurity.
2. Inconstancy; Events persist. Persistence involves some statistical inconstancy.

When the existing situation changes, events vary for stabilization and stabilize for continuity. Events have variability, continuity and substantiality.

For the variability the probabilities of existence and persistence are proportional and for the continuity the probabilities of existence and persistence are inversely proportional. Variability and continuity substantialize events with or without mass. Those events with low continuity will change into high continuity reducing variability to exist easier in favorable conditions.

These concepts stated above explain that stabilization principle is consistent with fundamental physics laws. The universal gravity is derived from the axiom of insecurity and the theory of special relativity is derived from the axiom of inconstancy. This principle deduces various natural propositions including variability, continuity and substantiality, which explain existence and evolution of events. Stabilization principle is summarized as follows. Events have existence probability and persistence probability. They induce the proposition of variability and continuity which coexist in substantiality. This principle is expected to be applied to various events and to various studies. The concept of variability and continuity should be acknowledged and be applied to various subjects. These statistical propositions of stabilizing events have been realized in the biological studies for formulating mathematical growth models of cell increase and tumor growth. The proposition is easily applicable to solve various events including rudimentary physics.

Contents

A. The essentiality of existence

1. The concepts of space and existence

(1) Space

It is mental perception in physics what actual space is. Spaces which we think of can be from one- to multi-dimensional. The high dimensional spaces are applied to the analyses of natural or social phenomena. They are conceptual spaces. The points in the conceptual space have exact positioning with perfect steadiness. To express the steadiness each space has to have a time axis.

The existence of objects is usually considered in the three-dimension space which has been named Euclidean space. The space which is an aggregate of points is determined with an origin and three fundamental directions. And the origin point has position and motion with complete stability in conceptual or subjective space. The existence of objective or actual space must be recognized. Actual space of great universe is hardly perceptible. Whatever is the space of great universe, the objective space of our vicinity has been perceptible as the three-dimension space with substantiality which is due to spatial and chrono-logical instability. Objective space is the space of temperature or heat, which is attributable to its substantiality. Objective space has position and motion which make some instability on reference to the subjective conceptual space. Hence the entire points in actual space have the instability of existence and persistence, which allow substantiality to entire actual space. Actual or physical space is sub-stantial existence.

Conceptual or physical space can have another secondary space in it. The reference frame for the secondary space is the primary space. The reference frame of primary physical space is conceptual space. Hence physical space has some instability of position and motion and has in consequence substantiality.

The physical space which can be designated here substantial space has always three dimensions with its own time.

The existence of objects is usually considered in the three-dimensional space with its time, therefore, the concept of time is required. The time considered in a space is elapsed time. The elapsed time which is called here time is the duration from one moment to another. Substantial spaces have always their own time. Several other secondary spaces are considered to exist in one primary space. Each substantial space has its own time whose value is also statistic. But in conceptual spaces time is considered communal in all spaces with scholar scale. In physical spaces the relativity of time occurs in dynamic events. Time is an independent variable existing in a space. Distance in a fundamental direction, for example, can become dependent variable of time. Space and time construct a frame of existence. Space and time are required to express the existence of a physical event. The existence of a mass point can be expressed in a space-time frame which is the substantial space the mass point belongs to.

(2) Existence

A point in a Euclidean space is considered to have position and motion. The position of a point has relativity to another point and the motion is the change of this relativity. Space, points, movements are stable. The position of a point changes, then its relativity, for example, to the origin changes. The above statement is the conceptual perception of existence. This concept of position and motion is employed in the perception of actual point locations which should have in nature some instability.

When there is an actual shape in a space, the actual position of a point on the shape coincides not always on the corresponding point of its virtual shape in space with perfect precision and with no instability. When the point is not conceptual but substantial, it does not always coincide perfectly to the corresponding point of the reference frame. It can shear from the corresponding point of the reference frame. The coincidence is mental and is essentially not perfect.

Its exactitude is statistical. The coincidence is not secured. Shears or deviations accompany to the coincidence. Shear or deviation decreases the statistical exactitude, which is existential insecurity. The expected position which is usually the center of existential probability distribution has the higher reliability. When a substantial point would not be in the center, it would move to the position with higher reliability which resides usually at the center. The reliability difference causes the centripetal force, which is stabilization and is the source of gravity. The space defines the existence and the time defines the change in substantial space. The physical objects in substantial space have position and motion collating with the existing substantial space, which induces dispersion and fluctuation and, therefore, they have insecurity and inconstancy. Existence varies because of time. Space makes insecurity. Time makes inconstancy.

A line is a range of points. Points, lines and shapes are defined as a point or an aggregation of points and, therefore, can be expressed in a space with three dimensions. A solid object, which is solidified energy or substance, is also considered as an aggregation of substantial points, therefore, its presence can be expressed in a reference frame. The presence of a substantial object in a substantial space does not necessarily coincide to the human conceptual presence. The human thought or conception is a scale and its quantity is scholar. The substantial quantity measured with this scale is statistic. An aggregation of points with insecurity in this meaning is a physical object or substance. Substantial objects have position and motion, and they have relative difficulties in the persistence of motion. They are called mass. Mass is designated to the relative difficulty in changing the motion velocity not depending on its size and sort. Essentially persistence probability represents the difficulty difference.

A physical object is often represented by one geometrical point neglecting its shape and size. This geometrical point represents the mass of object, which is the mass point. Presence of energy is often represented by one geometrical point and the physical quantity of energy is given to this point neglecting its shape and size which are hardly recognized.

2. Stabilization principle

Events exist and alter. Every event has primary quantity and secondary quantity with at least one communal parameter which causes statistical instabilities of both quantities. About electric quantities, for example, the electric charge is primary quantity which is determined by voltage but can alter. The alteration of this charge is electric current which is secondary quantity and is determined by voltage. Communal parameter is voltage. The secondary quantity is the alteration of the primary quantity. Another example is a mass point. A mass point resides at one point, which is primary quantity and is determined by distance. This existence has possibility to alter from the expected position, which is movement. This movement is secondary quantity and determined by quantities including distance. Distance is the communal parameter and causes the instability of existence and movement of a mass point. It is conceptional that the existence of the mass point is always exactly at an expected point. There should be deviations from the expected existence point. The deviation is statistic. The existence reliability changes according to deviations.

Every existence is hardly perfect. The reliability of an existence with little deviation may be high but not perfect. The existential probability at large deviation is low. The probability at small deviation is comparatively high. The probability distribution of the existence is normal distribution centering no deviation.

In every event the existence persists or alters. The alteration is the secondary quantity. The alteration of existence is also statistic. The probability of rapid alteration should be low. Probability of low rapidity is comparative high. Probability of no alteration which is persistence is high but not 100%. The persistence probability distribution is normal distribution centering no alteration.

Events have two major factors which are existence and persistence. About the example of mass point the existence of a mass point has position and motion. Motion is the persistence of position, which is obtained differentiating the distance with a parameter of time. Position has the existential probability distribution and motion has persistence probability distribution. Existence has insecurity and persistence has inconstancy. These two quantities are statistic. Events have two major factors which are existence and persistence and have statistic values of insecurity and inconstancy.

This principle of stabilization is constituted from the following two axioms of existence and persistence.

1. Insecurity; Events exist. Existence involves some statistical insecurity.
2. Inconstancy; Events persist. Persistence involves some statistical inconstancy.

3. Existential insecurity and inconstancy

A point which a man thinks about is a conceptual point which can be placed exactly at a position in a conceptual space and can make free movement. The point stays at a position in a conceptual space with perfect accuracy and its movement has complete steadiness. A substantial point based on nature is mainly a mass point and can be considered as the simplest existence. It does not stay or move exactly as a man thinks about. It is the existence which does not concern on human conception and follows natural laws. Conceptual existence is artificial. But physical existence which follows natural laws is involved with contingency. A point of an object exists with certain insecurity and is independent from human thought. In a physical space there is essential difference between the existence of the conceptual point and the existence of a point of a physical object which has shears and shakes. A mass point has position and motion. The position can comprise gaps and deviations, which are designated insecurity. And motion can have fluctuation, which is designated inconstancy.

B. Existential insecurity

(1) Dispersion

The sensuous evaluation of existence is made here. A man places a golf ball on a flat surface. A set of coordinates is made on the surface. The gravity center point of a golf ball is placed at a specific point, for example, at origin. The gravity center can be placed at origin, which is although conceptual. Errors are always involved in the reality. When this trial is repeated 100 times, it is difficult to think that the center point is always placed on origin. There should be dispersion. If college students perform this trial, the modal point may be the origin and the dispersion may be relatively small. If primary school boys perform this trial, the modal point may be also the origin. But the dispersion becomes larger. In both trials the gravity center point is placed at the origin but the nature of existence is different. The location is represented by the origin point but the dispersion accompanies with various standard deviation. The former trial has the smaller mean deviation and the higher precision where the locations are gathered in smaller vicinity. The latter trial has the lower precision with larger errors and lower security. There is difference in the existential precision and the latter trial has larger scattering and larger standard deviation value.

(2) Existence on a point

The existence is evaluated from the other direction. The gravity center of a golf ball is located at origin. Let the center represent the golf ball and is considered as the mass point which exists at origin. Conceptually the mass point is exactly secured at origin. The mass point exists very likely at the origin but there involves some insecurity. There could be dispersion. There could involve some errors. The reliability in the existential probability distribution of the mass point is the maximum at no dispersion or at the expected location which is the origin point and around it there lie lower probability densities. This situation is involved with high or low precision of the existence and is designated insecurity.

This situation is easily conceivable with particle locations. When the position of electrons or other particles at one instance is considered, the position can be calculated. But not all particles are at the calculated position. The value comprises some statistical dispersion which is insecurity. The calculated position is the modal value. Some particles are scattered around it. It is statistical existence.

(3) Existential probability distribution

Existential probability distribution indicates the possible existence at the expected point and its surroundings. The deviation from the expected point is the stochastic variable. The expected position is the modal value. It is the mean value of the variable in the distribution which is a normal distribution. The reliability at the modal value is the existential precision. The higher the value of this reliability is, the more stable the existence is. The value of standard deviation defines insecurity. The smaller the value of the standard deviation is, the more stable the existence is. The precision is defined by the reciprocal of standard deviation. To make it simple, the distribution of the x direction is considered. The occurrence on axis x is all over the same. Hence the probability distribution is the normal distribution (Φ) with a standard deviation (σ) and the expected point which is ($x = a$) in the center.

$$(1) \qquad \Phi = \exp[-(x-a)^2/(2\sigma^2)]/[\sigma(2\pi)^{1/2}]$$

σ is standard deviation of Φ $\quad \sigma > 0$

X shall be deviation on axis x from the expected point (a), then $X = x - a$.

$$(2) \qquad \Phi = \exp[-X^2/(2\sigma^2)]/[\sigma(2\pi)^{1/2}]$$

Reliability of the expected point ($X = 0$) is $1/[\sigma(2\pi)^{1/2}]$. The higher this value is, the steeper the reliability decreases, therefore, small gaps from (a) hardly exist with high reliability, which is precise existence. Reliability decreases as standard deviation (σ) increases, therefore, the magnitude of (σ) represents insecurity and its reciprocal ($1/\sigma$) expresses existential precision.

(4) Occurrence and insecurity

Let E be the energy at the expected point and let P_C be the occurrence of a unit of energy. Then the relationship between standard deviation (σ) of existence probability distribution and the occurrence (P_C) of unit of energy is the following.

(3) $\qquad \sigma^2 = P_C (1 - P_C)/E$

P_C is small enough compared to 1.

(4) $\qquad \sigma^2 \fallingdotseq P_C/E$

The standard deviation is the square root of the quotient of occurrence and energy. Mean variance is the same to the occurrence of a unit of energy or the occurrence rate. Hence the square root of energy and the reliability of existence probability are proportional. And the reliability and the square root of occurrence are inversely proportional. A mass point in the space of uniform occurrence has all the same mean variance of existence probability at any point in the space. A mass point in that space is stable and has no motion. But the reliability of existence probability is not 100%. The possibility to change its position coexists. Should it change the position, the existential reliability does not change. The same centripetal force works. The centripetal force around the mass point is uniform and the mass point exists stable. If occurrence would be non-uniform in the space and point (a) in the space would have the lower occurrence which is the higher reliability of existence probability. The point (a) has the higher existential reliability and the centripetal force works toward point (a). The mass point staying at point (b) which is the deviation X of point (a) moves toward point (a). The mass point obtains the higher existential reliability and the higher stability.

(5) Axiom of existence

Events exist. Existence is statistic quantity and involves some statistical instability, which is insecurity.

When a mass point is expected at point $x = a$, its deviation (X) is $(x-a)$.

A point at deviation (X) has the existential probability which is expressed with normal distribution (Φ).

$$(2) \qquad \Phi = \exp[-X^2/(2\sigma^2)]/[\sigma(2\pi)^{1/2}]$$

Existential precision is not perfect. Existence always comprises some insecurity which is expressed with standard deviation (σ).

C. Existential inconstancy

(1) Persistence

Among the existences of the same precision, some mass points easily change their positions and some points hardly change. Sensuous consideration on quivering is made here. A ping pong ball shall be placed at the origin of a set of coordinates. College students perform this action 100 times. The precision was as high as in the action with golf ball. But the stability is not as high as with golf ball. Ping pong ball is easy to move, which means it is weakly fixed. A golf ball stays still better. A golf ball has better impulse for standstill. A golf ball has high probability to have persistence. The existence with the same precision can have different steadiness or persistence. Now the weight center of a golf ball exists still at the origin point, which means there is no quivering or fluctuation speed is 0. That is the existence in which the steadfast situation has the maximum persistence. The higher the reliability is, the stronger fixture to the position is. The persistence possibility of a ping pong ball is much less compared with a golf ball and its steadfast situation has rather less fixture to the position.

(2) Fluctuation

When the existence on a point is considered, it has not only displacement possibility but also fluctuation possibility. An existence has essentially two elements of instability which are existential insecurity and inconstancy. Existence has not only positional instability whether a mass point is on the expected point, which is insecurity, but also qualitative instability of persistent motion. The latter instability is designated here inconstancy. Steady motion with high persistence has little motion fluctuation. A steadier golf ball has less fluctuation possibility or lower inconstancy. There are existences which are easy to fluctuate with high inconstancy or those which are hard to fluctuate with low inconstancy. For example, a heavy ball is an existence with low inconstancy and has relatively strong persistence and therefore, hardly fluctuates. A ping pong ball has low persistence and easily fluctuates. This difference is expressed by the difference of inconstancy. The low inconstancy means that the mass point has strong fixture to the position, which is high persistence. The same discussion can be made to a shape which consists of points. In the strong shape, which is an existence of low inconstancy, the positional relationship of representative plural points on a shape is hard to change.

Existence generally implies the two elements of precision and consistency, which complies that a mass point at a point has not only deviation but also fluctuation.

(3) Persistence probability distribution

A mass point has two elements which are position and motion. Motion instability distributes with normal distribution centering no fluctuation. Its standard deviation expresses the magnitude of the inconstancy. The stochastic variable of this persistency probability distribution is the time dependent change rate of the stochastic variable of existence probability distribution. The instability of this change rate is designated here fluctuation velocity or fluctuation. A standstill mass point has fluctuation modal value of 0 and its reliability is the standstill possibility. The probability distribution is a normal distribution and the smaller its standard deviation is, the higher the standstill possibility is, and more persistent. Hence the reciprocal of the standard deviation is designated here persistency. When a mass point is standstill at a point (a) in a Euclidean space, then the motion velocity is 0. There is possible fluctuation even in the standstill situation. At a point in a uniform space a mass stays still, then the mass point has uniform fluctuation possibility around it at any direction. In consequence, its mean fluctuation is 0.

Let X be the deviation of a mass point staying at point (a) on axis x, then the fluctuation velocity (V) is as follows.

$$(5) \qquad V = dX/dt$$

Fluctuation is the time dependent change rate of deviation and has the dimension of velocity. The mass point has fluctuation probability of normal distribution centering 0 on any direction. Let (Ψ) be the persistence probability distribution, then its standard deviation (τ) expresses the inconstancy and its reciprocal is the consistency or strength of persistence. Considering on axis x, a mass point staying still at ($x = a$) has a normal distribution of inconstancy centering 0 of fluctuation velocity (V).

$$(6) \qquad \Psi = \exp[-V^2/(2\tau^2)]/[\tau(2\pi)^{1/2}]$$

Reliability at $V = 0$ is $1/[\tau(2\pi)^{1/2}]$, which is the steadfast reliability. The higher this value is, the more abrupt fall the reliability with a small shift makes.

(4) Axiom of persistence

Events persist. Persistence is statistic quantity and involves some statistical instability, which is inconstancy.

A point at deviation (X) has the persistence probability which is expressed with normal distribution (Ψ) as equation (6). Persistence is not perfect. Persistence always comprises some inconstancy which is expressed with standard deviation (τ).

D. Actuality of mass points

(1) Variability of a mass point

A conceptual point can move freely in a conceptual space. But physical points or mass points cannot move freely in the space. The motion is the time dependent. Besides the existence probability restricts the motions. Particularly the relativity change of each point on a shape to the origin of a substantial space is restricted. Actual motion depends not only on persistence probability but also existence probability. Considering on motion, the concept of duration time is required. The elapsed time which is called here time is the duration from one moment to another. Motion changes the existence reliability and stabilizes the mass point, which is variability. The variability comprises the elapse of time. A mass point with high inconstancy is apt to change and together with high insecurity it becomes more changeable. For variability insecurity and inconstancy are in the same direction. Insecurity and inconstancy are proportional in variability. Variability is defined as the quotient of existence probability and persistence probability. As described later, the variability (Z) is a constant which is independent to the elapse of time.

(7) $\qquad Z = \Phi/\Psi$

(2) Continuity of a mass point

The substantial stability consists of variability and continuity. Natural events carry on or change. Changes are essentially for stability which is continuity. When situations change, events vary for continuity. For continuity events with high insecurity have to have low inconstancy and events with high inconstancy have to have low insecurity. In continuity of events insecurity and inconstancy are in opposite direction. Hence insecurity and inconstancy are inversely proportional in continuity. Persistence reliability and existence reliability are inversely proportional. Continuity is defined as the product of existence probability and persistence probability. As described later, the continuity (H) is a constant factor.

$$(8) \qquad H = \Phi * \Psi$$

(3) Substantiality of a mass point

The actuality of events is usually perceived with scholar values. But the values should be considered as the most probable statistic values which are determined by the balance of variability and continuity. When an event is in changing situation, the substantiality has large variability and its continuity should be small. And when it is in stabilized condition, its variability is small and its continuity is large. Variability and continuity are in opposite direction. They are inversely proportional in substantiality of events. Usual human perception often makes the following statement.

"Fragile things are hard to exist, which can, however, exist in favorable situation."

This expression can be explained as follows. Events with low continuity easily change having high variability but in situations with low variability, they have high continuity. Events with high continuity have low variability. Hence the substantiality (Θ) is defined as the product of variability and continuity. Substantiality (Θ) is the characteristic of an event and is its essentiality. Substantiality (Θ) is a constant.

$$(9) \qquad Z * H = \Theta$$

In the substantiality (Θ) of an event variability (Z) and continuity (H) are inversely proportional, which is expressed with above equation (9).

Logarithm of both sides of equation (9) is taken,

Log(Z)+log(H) = log(Θ)

And both sides are differentiated. Θ is a constant since substantiality is the characteristics of events. Hence $d\Theta/dt/\Theta$ is 0.

(10) $dZ/dt/Z+dH/dt/H = 0$

When the surrounding conditions do not change, continuity (H) does not change. Hence $dH/dt/H$ is 0. In consequence, $dZ/dt/Z$ is also 0. H and Z are constants under constant conditions.

E. Revolution and stabilization

1. Revolution of events

(1) Variability and revolution

Existence reliability and persistence reliability are proportional in variability. Hence the variability is the quotient of existence probability (Φ) and persistence probability (Ψ).

Variability (Z) is the proportional constant as shown below.

(7) $\Phi/\Psi = Z$

Logarithm of both sides are taken and differentiated,

(11) $(d\Phi/dt)/\Phi - (d\Psi/dt)/\Psi = (dZ/dt)/Z$

The variability induces that the remainder of insecurity variation rate and inconstancy variation rate is variability variation rate. Equations (2) and (6) are substituted into the equation above, then the equation below is obtained.

(12) $(dX/dt)X/\sigma^2 - (dV/dt)V/\tau^2 = (dZ/dt)/Z$

X is deviation from point (a) on axis x and V = dX/dt. Z is a constant.

Then the equation below is obtained.

(13) $d^2X/dt^2/\tau^2 - X/\sigma^2 = 0$

Let $\tau/\sigma = \omega$, then the dimension of ω is T^{-1}, which is the dimension of angular velocity.

(14) $d^2X/dt^2 = \omega^2 X$

The mass point at $x = a$ is expected to be revolving and has positive acceleration which causes revolving centrifugal force.

(2) Continuity and revolution

Continuity is the stabilized existence. The continuity (H) is the product of existence probability (Φ) and persistence probability (Ψ). The continuity (H) is the proportional constant. When a mass point (A) stays still at point (a) on axis x, the continuity is,

(8) $\Phi * \Psi = H$

Logarithm of the both sides of the above equation are taken and are differentiated.

(15) $(d\Phi/dt)/\Phi + (d\Psi/dt)/\Psi = dH/dt/H$

Equations (2) and (6) are substituted into the equation above, and then the equation below is obtained.

(16) $(dX/dt)X/\sigma^2 + (dV/dt)V/\tau^2 = dH/dt/H$

X is deviation from point (a) on axis x. $V = dX/dt$ and H is a constant. Hence the equation above becomes the equation below.

(17) $d^2X/dt^2/\tau^2 + X/\sigma^2 = 0$

The ratio of τ/σ is revolving angular velocity (ω). The equation above is an oscillating function. The mass point is revolving at the point (a). Then the equation below is obtained.

(18) $d^2X/dt^2 = -\omega^2X$

Continuity is waving. The mass point at (a) is expected to be revolving and has negative acceleration which causes revolving centripetal force. Centrifugal force by equation (14) and centripetal force by equation (18) keep the mass point stable. The stabilized mass point still has the angle velocity, which means it is revolving on point (a). As equation (14) indicates, the revolving mass point at point (a) has positive acceleration which causes revolving centrifugal force. And as equation (18) indicates, the mass point has negative acceleration which causes revolving centripetal force. The both forces stabilize the mass point. Events actualize a stabilized substantial mass point which is revolving.

2. Stabilization of mass points

When there would be a mass point at a point (b) in axis x whose existence reliability would be lower than that of point (a), then the mass point makes motion to point (a) on axis x. Point (b) is regarded as deviation (X_0) of point (a). Centripetal force from point (a) affects to the mass point at point (b). When the mass point is moving along axis x, above equation (13) can be changed as follows.

(19) $[d/dt+(\tau/\sigma)][d/dt-(\tau/\sigma)]X = 0$

(20) $dX/dt-(\tau/\sigma)X = 0$

Since equation (20) diverges to infinite, which is a destructive process and is not realistic. This is a stabilizing process and here $[dX/dt+(\tau/\sigma)X = 0]$ should be applied. The solution of equation (19) shall be as follows.

(21) $dX/dt+(\tau/\sigma)X = 0$

The solution of this equation is the following abducting function where X_0 is initial value of deviation X which is the position of point (b).

(22) $X = X_0*\exp[-(\tau/\sigma)t]$

The deviation reduces with the attenuation coefficient (τ/σ) and converges to 0, which means that the amplitude of wave function (18) converges to 0 as described next. But this is not realistic solution since continuity has to work on it.

The mass point makes motion according to the variability which is equation (21) and stabilizes according to the continuity which is equation (17). According to the substantiality, the sum of these two equations shows the motion.

Equation (21) shows the motion from point (b) to point (a) due to the variability. Equation (21) can be changed as follows.

(23) $2\omega dX/dt + 2\omega^2 X = 0$

The continuity contributes to the motion. Equation (18) is changed as below.

(24) $d^2X/dt^2 + \omega^2 X = 0$

According to equation (10), the substantiality of motion is the sum of equations (23) and (24), which is the equation below.

(25) $d^2X/dt^2 + 2\omega dX/dt + 3\omega^2 X = 0$

The solution of the equation above is as follows.

(26) $X = \exp(-\omega t) * [X_1 * \cos(2^{1/2}\omega t) + iX_2 * \sin(2^{1/2}\omega t)]$

X has to be a real number, therefore, $X_2 = 0$.

When $t = 0$, X is initial deviation X_0, therefore, $X_1 = X_0$.

Then equation (27) is obtained which is an abducting wave function.

(27) $X = X_0\exp(-\omega t)\cos(2^{1/2}\omega t)$

Wave amplitude X reduces and becomes close to 0 and the mass point is stabilized.

Equations (14) and (18) indicate a mass point has the material wave by De Broglie. These equations make a wave function which is material wave whose wave length λ should be $\lambda = h/P$. Light is wave and has also characteristics of particles. Then particles should have characteristics of wave. The wavelength of an electron is less than one thousandth of that of visible light. Particles and mass points have latent waving. Material wave of a moving particle has progressive waves.

3. Stabilization of spaces

A substantial space exists in the conceptual space or in a reference frame. The points in a substantial space have no mass. But they have positions with various deviations and motion with various fluctuations referring to corresponding points in conceptual space. Time in a substantial space makes existences change. Point in the space have small instabilities, therefore, it has existential probability and persistence probability, which induce continuity and variability. Hence it has substantiality with no mass. The substantial space consists of this kind of points, therefore, it is substantial and has instability. Variability of substantial points follows equation (13) and makes centrifugal acceleration. Continuity follows equation (18) and makes centripetal acceleration. Both equations make the stable existence of a space. The points in it are stable and are revolving. Hence the points in a substantial space are stable, therefore, a substantial space is stable.

F. Rotation and occurrences

1. Linear and spatial occurrence

Occurrence of energy or a mass is uniform in a physical space. Hence the occurrence on a line in the space is also uniform. In consequence, existence probability of a mass point forms a normal distribution. Let P_C be linear occurrence of a unit of energy and let E be amount of energy of a mass point. Its existence probability is a normal distribution and the relationship of standard deviation (insecurity) and occurrence is shown with equation (4) which is $\sigma^2 \fallingdotseq P_C/E$. Hence the square root of the quotient of linear occurrence and energy is the standard deviation which is the insecurity of a mass point. Occurrence of a mass point in a space is all the same at any point. Let ε^2 be the standard deviation of existential probability spatial distribution of energy E. Then the occurrence of energy E in a unit volume is $E*\varepsilon^2$. And its relationship to linear occurrence (P_C) is as follows, where Δs is the cross section in the vicinity of axis x.

$$(28) \qquad P_C*dx = \Delta s*E*\varepsilon^{2}*dx$$

$$E*\sigma^2 = \Delta s*E*\varepsilon^2$$

$$\sigma^2 = \Delta s*\varepsilon^2$$

$$(29) \qquad \Delta s = \sigma^2/\varepsilon^2$$

Dimension of Δs is area and dimension of σ^2 is square of distance, therefore, ε has null dimension.

2. Occurrence of another mass point

Occurrence of mass point (A) is all the same at any point in a space.

Hence linear occurrence (P_C) is uniform at any point over axis x. But when there is another mass point at origin, the appearance rate of mass point A on axis x is not uniform over the axis. Mass point A stays at $(x, 0)$. The appearance rate of origin point from mass point A is the configuration factor. The angle is regarded as proportional to the oscillation angular velocity (ω) and the solid angle is regarded now as a circular cone. The factor R_O is the ratio of base surface $\pi(a\omega x/2)^2$ and spherical surface $(4\pi x^2)$.

$$(30) \qquad R_O = \pi(a\omega x/2)^2/(4\pi x^2) \qquad a \text{ is proportional constant}$$

$$= (a\omega)^2/16$$

R_O is constant to x. The appearance of mass point A from origin R_{OX} is proportional to quotient of cross section Δs and spherical surface. Δs is replaced with equation (29).

$$(31) \qquad R_{OX} = b\Delta s/(4\pi x^2) \qquad b \text{ is proportional constant}$$

$$= (b\sigma^2)/(\varepsilon^2 * 4\pi x^2)$$

Hence occurrence (R_X) of mass point (A) at point $(x, 0)$ is as follows.

$$R_X = R_O * R_{OX}$$

$$= (a^2\omega^2/16)*(b\sigma^2)/(\varepsilon^2 * 4\pi x^2)$$

$$= (a^2\omega^2 * b\sigma^2)/(G * x^2) \quad \text{where } G = (8\varepsilon)^2\pi/(a^2 b)$$

$$(32) \qquad R_X = \tau^2/(G * x^2)$$

Occurrence of mass point A at point $(x, 0)$ is proportional to τ^2 and inversely proportional to x^2.

3. Influence to another mass point

When a mass point A exists in a space, it can stay at any point in the space since occurrence is uniform. Now a mass point A stays at point $(x, 0)$ which is stable. If another mass point would exist in the space, the mass point A is no more stable. They influence each other. The existence of mass point A at point $(x, 0)$ is not stable. The influence of mass point A to the mass point at origin is considered here. The more distance from origin it has, the more stable it is. The occurrence of mass point A at point $(x, 0)$ in relation to the mass point at origin is inversely proportional to the square of distance x as expressed with equation (32). A stochastic variable u is postulated and u is defined as the equation below.

(33) $u^2 = 1/x$

When the value of u is 0, x has to be infinite. While value x increases from x to infinite, value u has to decrease from u to 0. Meantime the integrated occurrences of both variables have to be the same. The occurrence rate decreases with x, which is shown with equation (32). The decrease at x is $(R_x/x^2)*dx$ which is integrated. Then the result is $-\tau^2/(2G*x)$ which should be equal to $-(\tau^2 u^2)/(2G)$. Hence the occurrence rate of u is expected constant, which has to be verified. The occurrence rate of the mass point A is assumed to be constant at any value of u. Let q be the occurrence rate of u at any value of u. Then $q*u*du$ is the increase. It should be the same with the decrease of occurrence rate with x which is dx.

(34) $-R_x*dx = q*u*du$

Both sides are separately integrated. x is from infinite to x and u is from 0 to u.

(35) $\tau^2/(2G*x) = q*u^2/2$

When $u^2 = 1/x$, the integrated occurrences are the same, therefore, $q = \tau^2/G$.

Then mass point (A) makes normal distribution with variable (u). The mass point (A) is the most stable when x is infinite, which means when u is 0. Occurrence is constant at any value of (u), which means the mean of (u) value is 0. Existence reliability of mass point (A) is the maximum at $u = 0$. Hence existence probability distribution with stochastic variable (u) forms normal distribution and is as follows.

(36) $\Phi = \exp[-u^2/(2\eta^2)]/[\eta(2\pi)^{1/2}]$

η is standard deviation. Equation (35) is substituted into this equation. Then equation (37) is obtained.

(37) $\Phi = \exp[-1/(2x\eta^2)]/[\eta(2\pi)^{1/2}]$

This is the existence probability distribution on axis x. Persistence distribution is as equation (6). It is a normal distribution centering 0.

(6) $\Psi = \exp[-V^2/(2\tau^2)]/[\tau(2\pi)^{1/2}]$

The continuity is applied to these two equations. Then the equation below is obtained.

(38) $d^2x/(dt)^2 = -(\tau^2/\eta^2)/x^2$

Equation (38) shows that a mass point at point (x, 0) have the centripetal acceleration toward the origin as gravity which is inversely proportional to square of distance (x).

When $x = 1$,

From equation (32) $R_1 = \tau^2/G$

From equation (4) $\eta^2 = R_1/E$

$\eta^2 = \tau^2/(G*E)$

The above equation is substituted into equation (38).

(39) $-d^2x/dt^2 = (G*E)/(x^2)$

Hence a mass point at $(x, 0)$ has the acceleration toward origin which is proportional to energy and inversely proportional to square of distance as the universal gravity.

4. Rotation

Continuity makes centripetal force as equation (38) describes. The quotient of inconstancy and insecurity of equation (38) is angle velocity which means the mass point at (a) rotates around the rotation center.

Variability should be applied to equations (37) and (6). The resulting equation is as follows and describes the centrifugal acceleration.

(40) $\qquad d^2x/dt^2 = (\tau^2/\eta^2)/x^2$

The positive acceleration causes centrifugal force. Continuity provides the centripetal force of the same quantity as described in equation (38), therefore, the distance between the mass point and the rotation center does not change in rotation.

G. Gravity

1. Restoration force

Suppose a mass point now stays still at origin, and its standard deviation of existence probability is small, which means the mass point has high value of precision and high reliability. A small displacement makes a large difference of the reliability. And the mass point has high possibility to recover the position right away. That means force works toward the expected point. When the mass point would have lower precision value, a small displacement would make small reliability decrease, therefore, the restoration force is small. The existence with low insecurity means that the restoration force is relatively large and the mass point has the higher probability to stay at the expected point. This condition has high precision and small existential insecurity. The restoration force which is centripetal force at the expected point is designated as central force.

2. Displacement and centripetal force

A stationary mass point has uniform centripetal force around itself. Therefore, it can keep the static condition. A small displacement decreases the existential reliability and the restoration force works since the center has higher existential reliability. The centripetal force is the strongest at the expected point and it decreases proportionally to reliability in the peripheral. Existence with high precision has strong centripetal force and hardly makes deviation. Existence with low precision possibly makes deviation since central force is not very strong. It is proportional to reliability at the center, which is the expected existential point. And it is proportional to the existential precision ($1/\sigma$). From equation (2) the central force (F) is proportional to $1/[\sigma(2\pi)^{1/2}]$.

(41) $F = E/[\sigma(2\pi)^{1/2}]$

Proportional constant E has the dimension of energy and is considered as the mass point energy. The centripetal force, which causes the gravity, is a function of X. If there exists energy K at deviation X, the reliability of energy E causes centripetal force to energy K toward energy E. The centripetal force is the reliability at X multiplied by energy K. At the same time, energy K has the existential reliability at X = 0 which causes centripetal force to energy E toward K. The product of both centripetal forces makes the gravity.

3. The universal gravity

The existence of a unit of energy at the origin point has its existence probability distribution with insecurity (σ_1), which is $\exp[-X^2/(2\sigma_1{}^2)]/[\sigma_1(2\pi)^{1/2}]$ and causes centripetal force in its vicinity. A mass point of energy K at the deviation X receives the centripetal force toward X = 0. The centripetal force ($-F$) of energy K is as follows.

(42) $-F = K\exp[-X^2/(2\sigma_1{}^2)]/[(2\pi)^{1/2}\sigma_1]$

The energy incidence at X = 0 multiplies this centripetal force. When the energy E resides at X = 0, the existential probability distribution of energy K determines the existential incidence of energy E at X = 0, which is $E\exp[-(-X)^2/(2\sigma^2)]/[(2\pi)^{1/2}\sigma]$ and modifies the centripetal force of energy K at X.

The product of this energy incidence and equation (42) is the centripetal force ($-F$).

$$-F = E\exp[-(-X)^2/(2\sigma^2)]/[\sigma(2\pi)^{1/2}]*K\exp[-X^2/(2\sigma_1{}^2)]/[\sigma_1(2\pi)^{1/2}]$$

$$= KE\exp[-(-X)^2/(2\sigma^2)-X^2/(2\sigma_1{}^2)]/(2\pi\sigma_1\sigma)$$

$$= KE\exp[-X^2(\sigma^2+\sigma_1{}^2)/(2\sigma^2\sigma_1{}^2)]/(2\pi\sigma_1\sigma)$$

When X is large enough, the approximation is done applying Taylor expansion.

(43) $\quad -F = KE[2\sigma^2\sigma_1{}^2/X^2(\sigma^2+\sigma_1{}^2)]/(2\pi\sigma_1\sigma)^2$

$\quad\quad\quad = KE[\sigma\sigma_1/(\sigma^2+\sigma_1{}^2)]/(\pi X^2)$

Newton's equation of the universal gravity is obtained.

$\quad\quad -F = KE\{\sigma\sigma_1/[\pi(\sigma^2+\sigma_1{}^2)X^2]\}$

(44) $\quad -F = G*KE/X^2$

G is the gravitation constant which is $G = \{\sigma\sigma_1/[\pi(\sigma^2+\sigma_1{}^2)]\}$.

The gravity between energy E and energy K is the product of energy incidence of energy E at K point and energy incidence of energy K at E point.

H. Relativity

1. Momentum and energy

A mass point is at a point, which means it stays still and is stable at the point. Its motion velocity is 0. This velocity 0 has some instability. The mass point can fluctuate but it has the maximum reliability at fluctuation velocity 0 in its persistence probability distribution and the lower reliabilities exist around it. If it does deviate from the center, the reliability difference of persistence distribution makes the motion of mass point to return quickly to the center where the existence has high persistency reliability. It is, therefore, steady state. The standard deviation τ expresses the magnitude of inconstancy. The smaller it is, the stronger the existential persistency is. The reciprocal of this standard deviation which is $(1/\tau)$ is proportional to the existential persistency. Hence it is designated consistency and is proportional to momentum.

Now a mass point of energy (E_1) stays still $(v = 0)$ at point (a) on axis x, then the persistence reliability is $1/[\tau(2\pi)^{1/2}]$ applying equation (6). The product of energy E_1 of a mass point and persistency reliability has the dimension of momentum.

$$(45) \qquad P = E_1/[\tau(2\pi)^{1/2}]$$

Hence momentum (P) is inversely proportional to inconstancy of which proportional constant has the dimension of energy and is considered the energy of mass point.

Product of momentum and angular velocity is force. The quotient of central force (F) and static momentum (P) is angular velocity (ω). F/P is obtained from equations (41) and (45).

$$F/P = [E/(2\pi\sigma^2)^{1/2}]/[E_1/(2\pi\tau^2)^{1/2}]$$

$$= (\tau/\sigma)/(E_1/E)$$

(F/P) is (ω) and (τ/σ) is (ω), therefore, $E_1/E = 1$. E is communal to insecurity and to inconstancy, which is the energy and represents the mass point and defines its central force and momentum.

2. Velocity and relativity

Space α contains space β which contains a mass point. Its static energy is E and static momentum is $P_0 = E/(2\pi\tau^2)^{1/2}$ as equation (45) shows. When space β moves with velocity v in space α, the energy and momentum do not change in space β. They are still static. In space α momentum changes but static energy E does not change. The momentum of a mass point in motion in space α shall be obtained. Let (τ') be inconstancy in motion which is the standard deviation of the moving mass in its persistency probability distribution in the moving direction in space (α). From equation (45), the momentum in standstill (P_0) is $E/(2\pi\tau^2)^{1/2}$, and the momentum (P) in motion with velocity (v) shall be $E/(2\pi\tau'^2)^{1/2}$. Momentum (P) increases with the increase of velocity. Then inconstancy τ has to decrease to (τ') but static energy (E) does not change. A moving mass point is affected by its velocity. It changes the inconstancy (τ') which can be described with a dependent function of (v). The static fluctuation (V) is decreased in the moving direction. The decrease should be expressed as the product of the velocity and another positive function (f). Equation (46) shows the mean value of the decreased fluctuation which is (τ'). If there would be no effect from the velocity, function (f) should be 0. The square of (τ') is the mean of variation $(V-fv)^2$ which is the subtraction of square of the postulated mean inconstancy $(fv)^2$ from the square of standstill inconstancy which is (τ^2) of equation (6) when approximate fv is close to (τ).

(46) $(\tau')^2 \doteqdot \tau^2 - (fv)^2$

When the mass point is standstill ($v = 0$), $(\tau')^2$ equals to τ^2. As v is extrapolated upward, $(fv)^2$ increases, then $(\tau')^2$ decreases. But $(\tau')^2$ is 0 or more. Hence the maximum value of $(fv)^2$ is τ^2. Let C be the maximum value of v. Then, the following equations are obtained.

(47) $\tau^2 = (fC)^2$

(48) $f^2 = \tau^2/C^2$

(49) $(\tau')^2 = \tau^2[1-(v/C)^2]$

The velocity of a mass point has the limitation which should be the light speed or the speed close to it. It has been considered as the light speed in the theory of special relativity.

The momentum of a mass point in motion in space (α) is obtained. From equation (45), the momentum in standstill (P_0) is $E/(2\pi\tau^2)^{1/2}$, and the momentum (P) in motion with velocity (v) is $E/(2\pi\tau'^2)^{1/2}$, then the equation below is obtained.

(50) $P = P_0/[1-(v/C)^2]^{1/2}$

As momentum increases, the velocity increases and comes to close to the limitation which has been considered the light speed.

3. Mass and energy

A mass point in standstill has persistency distribution and maintains corresponding energy. A mass point in motion decreases its inconstancy and increases its energy (E) as motion velocity increases. The velocity has the limitation which is the maximum velocity (C). At the speed limitation energy increases no more.

The both sides of equation (45) are multiplied by C. CP has the dimension of energy.

(51) $CP = CE/[\tau(2\pi)^{1/2}]$

CP is the total energy (E).

(52) $E = CP$

Let E_0 be the static energy and E be the total energy of a moving mass point. From equation (52), $E_0 = P_0*C$. Equation (50) is changed as equation (53).

(53) $E = E_0/[1-(v/C)^2]^{1/2}$

The approximation of equation (53) is as follows.

$$E = E_0*[1+(v/C)^2/2]$$

$$= E_0+(P_0/C)v^2/2$$

P/C has the dimension of mass. The dynamic energy is $(P_0/C)v^2/2$ which should be equal to the dynamic energy by Newton $(m_0v^2/2)$.

$$E = E_0+m_0v^2/2$$

Hence, $m_0 = P_0/C$

Both sides of equation (50) are divided by C.

$$P/C = (P_0/C)/[1-(v/C)^2]^{1/2}$$

Then $m_0 = P_0/C$, m_0 is static mass and, therefore, $P/C = m$, m is the mass of moving mass point.

From equation (52), $E/C = P$, then $E/C^2 = m$.

Both sides of equation (53) is divided by C^2.

$$E/C^2 = (E_0/C^2)/[1-(v/C)^2]^{1/2}$$

The total energy divided by the square of the maximum velocity (E/C^2) is the total mass, therefore, following equation (54) is obtained.

(54) $m = m_0/[1-(v/C)^2]^{1/2}$

m is mass, m_0 is static mass

I. Uncertainty and variability

(1) Variability of momentum

Momentum (P) has insecurity and inconstancy. The derivative from the differentiation of momentum is force, therefore, the argument of momentum persistency probability distribution is force. Hence momentum and force compose continuity and variability of momentum. The variability of momentum is induced as follows.

The existence probability distribution of momentum deviation (P_X) is the normal distribution (Φ) with a standard deviation (σ_p) and with the expected momentum which is (P_0) in the center as is expressed with equation (55).

$$(55) \qquad \Phi = \exp[-(P-P_0)^2/(2\sigma_p{}^2)]/[\sigma_p(2\pi)^{1/2}]$$

σ_p is standard deviation of Φ

P_X shall be deviation from the expected momentum (P_0), then $P_X = P-P_0$.

$$(56) \qquad \Phi = \exp[-P_X{}^2/(2\sigma_p{}^2)]/[\sigma_p(2\pi)^{1/2}]$$

The fluctuation is the time dependent change rate of deviation (P_X) and has the dimension of force.

$$(57) \qquad F = dP_X/dt$$

This equation expresses the law of inertia when $F = 0$. The argument of momentum inconstancy probability distribution is the differential of momentum which is force.

Let (Ψ) be the persistence probability distribution.

$$(58) \qquad \Psi = \exp[-F^2/(2\tau_p{}^2)]/[\tau_p(2\pi)^{1/2}]$$

Since insecurity and inconstancy are proportional in variability, variability (Z) is the quotient of existence probability (Φ) and persistence probability (Ψ) as equation (7) shows. Equations (56) and (58) are substituted into equation (11) and processed and then equation (59) is obtained.

$$(59) \qquad (dP_X/dt)P/\sigma_p^2 - (dF/dt)F/\tau_p^2 = (dZ/dt)/Z$$

P_X is deviation from point (P_0) on axis p and $F = dP_X/dt$.

When Z is constant, the next equation is obtained.

$$d^2P_X/dt^2/\tau_p^2 - P_X/\sigma_p^2 = 0$$

$$[dP_X/dt + (\tau_p/\sigma_p)][dP_X/dt - (\tau_p/\sigma_p)]P_X = 0$$

Since $dP_X/dt - (\tau_p/\sigma_p)P_X = 0$ diverges, the solution is as follows where $\tau_p/\sigma_p = \omega_p$.

$$(60) \qquad dP_X/dt + \omega_p P_X = 0$$

This equation clarifies the action and reaction theorem. When there is an action momentum of P_X which is action force of $\omega_p P_X$, there occurs reaction force of $-dP_X/dt$.

(2) Theorem of uncertainty

Physical events always involve some dispersion and fluctuation. Probability of no dispersion or no fluctuation does not exist, which means events should involve some fluctuation. Insecurity with some dispersion and inconstancy with some fluctuation induces continuity and variability of events. Events alter. Uncertainty principle is usually expressed $\Delta E * \Delta t \geqq h$ or $\Delta P * \Delta x \geqq h$. Topological distribution of energy is force and the derivative from chronological differentiation of momentum is also force. $\Delta E / \Delta x \fallingdotseq F$ and $\Delta P / \Delta t \fallingdotseq F$, therefore, $\Delta E * \Delta t = \Delta P * \Delta x$. Both expressions are the same. Momentum is employed to elucidate this principle. Events have variability and continuity. Momentum also has variability and continuity. When momentum increases from P_0 to P, the increasing curve makes an abducting exponential curve, which is due to the variability.

The variability explains the uncertainty principle by Heisenberg.

(3) Uncertainty position and motion

Uncertainty principle is usually expressed as follows.

(61) $\Delta x * \Delta P \geq n * h \quad n \geq 1$

$\Delta x * \Delta P$ is differentiated.

$$d(\Delta x \Delta P)/dt = \Delta x * d\Delta P/dt + \Delta P * d\Delta x/dt$$

Variability is applied to Δx and ΔP. Δx can be regarded as the small displacement of x, and ΔP as the small displacement of P, and then $d\Delta x/dt = -\omega_1 \Delta x$ and $d\Delta P/dt = -\omega_2 \Delta P$ are obtained as equation (21&60) describes. Hence,

(62) $d(\Delta x \Delta P)/dt = -\omega_1 \Delta x \Delta P - \omega_2 \Delta x \Delta P$

$$= -(\omega_1 + \omega_2)\Delta x \Delta P$$

Let ω_3 be the sum of $\omega_1 + \omega_2$, then,

$$= -\omega_3 \Delta x \Delta P.$$

Δx and ΔP are positive quantities, therefore, $d(\Delta x \Delta P)/dt$ is a negative function and has the dimension of energy which is n time of energy quantum $h\omega$.

$$d(\Delta x \Delta P)/dt = -nh\omega$$

$$-\omega_3 \Delta x \Delta P = -nh\omega$$

ω and ω_3 are the same since they are the angular velocity of energy $d(\Delta x \Delta P)/dt$. And then,

$$\Delta x \Delta P = nh \quad n \geq 1$$

Hence,

(63) $\Delta x \Delta P \geq h$

(4) Uncertainty of energy and time

A quick change with certain energy loss requires tiny time (Δt) which cannot be zero. Even a lasting event accompanies slight energy change (ΔE) which cannot be zero. The product of ΔE and Δt is never zero, which has been known in quantum level as the uncertainty principle. This formula ($\Delta E * \Delta t \geq h$) is demonstrated as follows.

The differentiation of momentum is force ($dP/dt = F$). $\Delta P/\Delta t$ is considered here the approximation of dP/dt, therefore,

$$\Delta t \doteqdot \Delta P/F$$

Equation (45) is, $P = E/[\tau(2\pi)^{1/2}]$

$$\Delta P = \Delta E/[\tau(2\pi)^{1/2}]$$

Equation (41) is, $F = E/[\sigma(2\pi)^{1/2}]$

Therefore, $\Delta t = \Delta E/E/\omega$ $\omega = \tau/\sigma$

$$\Delta E \Delta t = (\Delta E/E) * \Delta E/\omega$$

When E is very small and E can be considered close to ΔE, then $\Delta E/E \doteqdot 1$.
$\Delta E = nh\omega$, therefore,

(64) $\Delta E \Delta t = nh$

Hence, $\Delta E \Delta t \geq h$

But when E is large,

$$\Delta E/E \doteqdot 0$$

Then,

$$\Delta E \Delta t \doteqdot 0$$

The proposition of uncertainty is applicable to very small mass points which may be the quantum level.

J. Hypothetic state of existence

1. Hypothetic state of energy

The standard deviation in equation (2) defines insecurity. As the standard deviation increases, the existence reliability of a mass point decreases and the substantiality of a mass point disappears with infinite value of insecurity. Equation (41) shows that the centripetal force disappears with infinite insecurity. This state is hypothetic. A stable flat distribution of the existence reliability with positive or negative infinite insecurity is conceivable. A mass point or energy becomes a flat density of existence probability which is represented by the distribution of energy E of equation (41). Hypothetic existence with a negative insecurity is also conceivable where centrifugal force occurs with equation (41) and the energy E represents the existence. And hypothetic energy excess (E) and energy deficits (−E) with negative insecurities can be considered in the hypothetic state. The centrifugal force occurs to an energy excess and the centripetal force occurs to an energy deficit with hypothetic negative insecurity.

2. Hypostatic forces

Equation (2) is applied to the hypothetic mass with negative reliability. X shall be deviation from the expected point of a hypothetic energy mass which is an energy excess or an energy deficit.

$$(2) \qquad \Phi = \exp[-X^2/(2\sigma^2)]/[\sigma(2\pi)^{1/2}]$$

The centrifugal force works to an energy excess. And the smaller absolute value of the standard deviation is, the stronger the centrifugal force is. The energy deficit draws energy and the deficits have tendency to restore. The excess has tendency to decay. When equation (41) is applied to the excess where F is negative and is centrifugal force. Frailty of the excess at the expected point (X = 0) is $1/[\sigma(2\pi)^{1/2}]$. The higher this absolute value is, the frailer the excess is, therefore, the excess has the more pressure to be flattened. Frailty increases as the absolute number of standard deviation (σ) decreases, therefore, the reciprocal ($1/\sigma$) expresses the tension to decay. The centrifugal force is proportional to the tension ($1/\sigma$). From equation (2) the central force (F) is proportional to $1/[\sigma(2\pi)^{1/2}]$.

(41) $F = E/[\sigma(2\pi)^{1/2}]$

Proportional constant E is considered as the excess energy. When E is a unit of energy, the probability distribution is the same to the excess energy incidence at deviation X. $-F$ is the tension in the center of distribution.

3. Substantial existence and hypostatic state

A particle is a substantial existence with mass which is consisted of energy. An example is a neutron which is a substantial existence. A neutron can become a proton and an electron which are consisted of energy. When a neutron releases an electron, the new existence which is a proton cannot always receive perfectly balanced energy. An energy excess which is a hypothetic existence can be left in a new proton. At the same time, the electron has an energy deficit which is also hypothetic existence. These proton and electron generate Coulomb force and the electron exists rotating around the proton with a certain angle velocity. High intensity energy which is energy of high frequency wave quantizes or particulates. Particles have their own proper energy levels, and energy excess or energy deficit can happen. These excess or deficit are hypostatic state of energy.

4. Coulomb force

Energy mound is fragile and causes centrifugal force in its vicinity. A mound of energy E which is considered as +Coulomb makes the centrifugal force. The energy incidence at the center of energy E is $E/[\sigma(2\pi)^{1/2}]$ and its centrifugal force is obtained applying equation (14). When there is another energy mound (K) at deviation X of the existence probability distribution of energy E, the existential incidence of energy E is modified due to the existence probability distribution of energy K. Energy K with +Coulomb modifies the centrifugal force of energy E. The centrifugal force of energy E at its center is given by the product of the energy incidences of energy E and K. In this case, σ is communal and F is centrifugal force in Coulomb force. Equation of the Coulomb force is obtained as follows.

$$F = E\exp[-0^2/(2\sigma^2)]/[\sigma(2\pi)^{1/2}]*K\exp[-X^2/(2\sigma^2)]/[\sigma(2\pi)^{1/2}]$$

$$= KE\exp[-X^2/(2\sigma^2)]/(2\pi\sigma^2)$$

When X is large enough, the approximation is done applying Taylor expansion as is done in equation (43).

$$F = KE[2\sigma^2/(-X)^2]/(2\pi\sigma^2)$$

$$= KE/(\pi X^2)$$

When the energy at X = 0 is an energy deficit (−E), the above equation expresses centripetal force, which is,

$$-F = KE/(\pi X^2)$$

Negative Coulomb draws positive Coulomb. Hence Coulomb (q) of energy (E) is,

$$(65) \qquad q = E/\pi^{1/2}$$

K. Energy waves

1. Space instability

The great space of universe has been hardly understood. Its existence is conceptual. Space is usually considered with multi dimensions, which is conceptual. But the actual space of three dimensions exists where various substances reside. A space which has one basic point and three fundamental directions gives secondary spaces which are substantial. They are movable, therefore, they are unstable. Hence they have variability and continuity, which means they have substantiality. Hence all points in actual spaces have the existence probability distribution with insecurity and the persistence probability distribution with inconstancy. There is clear distinction between a conceptual space with exact positioning and with no instability and an actual space which physically exists. Physical existence is usually recognized with existence of mass. But the existence with no mass has to be recognized.

Even extremely small spaces in an actual space have substantiality. A micro-space around a point which is an assemblage of points in an actual space has substantiality. This point represents the micro-space and is named substantial point. A point itself is conceptual but a substantial point represents the micro-space around it of which size is not determined. Equations (14) and (18) show all substantial points and their assemblages are rotating or waving. Substantial space is potential energy. Substantial space is the source of all substantial existences.

2. Energy source and waves

Space with time is potential energy. Physically existing points with various potentiality have position and motion. They also have some instability on which contingency works. Contingency can condense those points and make assemblages of points. These assemblages have their own existence probability and persistence probability distributions. They have various insecurities and inconstancies. Hence those assemblages have various substantiality which make them revolve with various intensity. Those revolving intensities depend on the size of assemblage which are incidentally made. The basis of energy is aggregated or condensed points which is the assemblage and is the source of energy. Large assemblage has high potentiality to yield high intensity energy. The intensity of these assemblage distributes with Poisson distribution. Majority is low intensity assemblage.

Our space has temperature which is energy and means our space is filled with low intensity energy. High intensity assemblage is not much. Those assemblages move around and become waves which are energy. Physical space is filled with energy waves of various intensity. High insecurity energy sources have tendency to disperse and high inconstancy energy sources have tendency to change their energy modes to become quanta or particles.

As equations (14) and (18) show the substantial space is consisted of revolving assemblages. High insecurity assemblages disperse easily. Low insecurity assemblages are not easy to disperse.

Due to insecurity and inconstancy these assemblages move. When they move, they form waves. As equation (13) indicates, assemblages with low frequency have low inconstancy and high insecurity, therefore, this kind of energy disperse easily, which is heat beam. Assemblages with high frequency have high inconstancy and low insecurity. They change easily and are not easy to disperse. They gather and increase the potentiality. They form assemblages of high potentiality and make energy waves of various high frequencies which form beaded sequences. They are energy quanta as light beams. Very high potential assemblages with very high frequency waves have very high inconstancy and very low insecurity, which means they hardly spread. They revolve and form very tiny beads which gather and particulate with or without mass. They are quarks and mesons. Mass is expected from strong spinnings.

The assemblages are energy source which repeat aggregation and extinction. Energy decline can happen. High intensity assemblage has low incidence in Poisson probability distribution. Hence it has tendency to decline down to lower intensity assemblage. Low intensity assemblage may extinct into assemblages. Total energy potentiality may scarcely alter.

3. Energy

The most essential existence is energy. Every point in substantial space has some instability collating it with the corresponding point in the conceptual space. Points in the substantial space have some dispersion and some fluctuation, and they are revolving which has been shown by equation (14) and (18). Revolving points make waves when they disperse. Wave is energy. Substantial space is potential energy. Energy sources in it define energy potentiality. High intensity energy is anticipated to require high potentiality. Occurrence of high intensity energy is expected to be lower than low intensity energy since the intensity distribution of energy sources is expected to be Poisson distribution. Energy has quantity and intensity. Quantity depends on the wave height. Large ocean waves have high quantity of energy. Small beach waves have low quantity of energy. Wave height depends on the amount of energy carriers. Intensity depends on frequency. Low intensity energy is heat. Waves from high insecurity points have tendency to spread, which is heat. Waves from high inconstancy points have high intensity and easily change their energy mode. Light quanta have higher intensity. High frequency waves move like a straight line. The highest intensity energy which has high inconstancy and high continuity just spins and forms mass. The high frequencies particulate and form quarks, mesons and elementary particles with or without mass. Light quanta and mass points are energy carriers and the carriers have position and motion and their own substantiality. Time is the most important factor in substantial space. Substantial space is filled with waves which are energy. Wave is energy.

4. Heat

Heat exists with insecurity and inconstancy. Hence heat follows the variability and continuity. The angular velocity ($\omega = \tau/\sigma$) should be limited. Heat is energy. Heat has waves of rather low frequency. Heat energy has generally rather high insecurity, therefore, its variability is small. It hardly changes. Heat exists localized in a substantial space. Heat is energy and is the most elemental existence. Heat is rather dispersive in space and its existential reliability is considerably low but never flat in space, therefore, the existential insecurity (σ) should be large but limited. Heat slowly infiltrates with an inconstancy (τ) into any direction and changes little. Hence its inconstancy (τ) is small. Energy is just a kind of wave. The wave function of equation (18) describes the frequency of heat. Heat should have rather low angle velocity. When angle velocity which is the energy intensity goes up more, then energy quantizes. And with very high angle velocity energy particulates and forms particles with mass which are substances.

5. Light

High frequency waves pile up and form beaded sequences which are light quanta. Light is a quantized existence which has its own substantiality. Light quanta are energy carriers consisted of energy waves. Light quanta have position and motion and their own existence probability and persistence probability distributions. Therefore, they have their own insecurity and inconstancy and have variability and continuity. Light disappear right away since light quanta have very high inconstancy (τ). They hardly last. Their persistence reliability is very low. They change into energy waves right away. The existential reliability of light quanta is high and their existential insecurity (σ) is small. Hence their angular velocity ($\omega = \tau/\sigma$) is large. Light quanta exist localized in space but only at one instance. Equation (18) describes its waving continuity which indicates light quanta have high frequency wave. Variability equation (22) shows they have little deviation. It also describes the variability is close to zero. Hence light quanta have almost no variability and move in a straight line in a space. Light goes straight and is affected little by anything around. Light quantum is integrated with energy waves which is material wave and confined in a straight line. Light quantum is an extreme substance. Light quantum has substantiality whose components have extreme values but still has the character of waves. Light is quantized energy and the energy wave which light quantum carries has the wave length of 4000–8000 Å, which is material wave. Quanta which have frequency above that of light quanta spin strongly and particulate. Particles with very high spinning are considered to generate mass.

L. Lorenz function

1. Lorenz function and variability

Lorenz function is induced from the variability. Equation (13) is derived from the variability, which is shown below.

(13) $d^2X/dt^2/\tau^2 - X/\sigma^2 = 0$

Both sides of equation (13) are integrated. C_O is a constant. τ' in the next equations below is the inconstancy of a moving space time frame and τ is the inconstancy of the fundamental space time frame.

$$\sigma^2 dX/dt - \tau'^2 X^2/2 = C_O'$$

$$\sigma^2 dX/dt - \tau^2 X^2/2 = C_O$$

The lower equation is subtracted by the upper equation.

$$\tau'^2 - \tau^2 = -2(C_O - C_O')/X^2 \quad \text{then,}$$

$$\tau'^2 = \tau^2 - 2(C_O' - C_O)/X^2$$

The dimension of both τ'^2 and τ^2 are the square of velocity, therefore, $2(C_O' - C_O)/X^2$ is square of velocity and is replaced with $(fv)^2$. f is a constant.

$$\tau'^2 = \tau^2 - f^2 v^2$$

As v increases, τ'^2 decreases. τ'^2 is 0 or more, therefore, when τ'^2 is 0, v is the maximum velocity which is C.

$$f^2 = \tau^2/C^2$$

$$\tau/\tau' = 1/(1 - v^2/C^2)^{1/2}$$

Right side of the equation above shows Lorenz function. It is necessary for revising mass, energy, momentum, etc., and is designated here revising γ function. The value of γ function is 1 or more and is required even for coordinates conversion.

2. Relativity of motion

The motion of a mass point is considered applying a coordinates system based on an origin point and a set of three coordinates. A space which based on an origin (R) is named here the coordinates system (α). Another space (β) of which origin point (R') is moving on the axis x of space (α) with velocity (v). A mass point (A) staying still with inconstancy (τ) at a point in the space (β) which is moving in space (α) in direction of axis x with inconstancy (τ') and velocity (v). Now inconstancy of a moving mass point in the space (α) is considered. The persistence probability distribution (Ψ) in space (β) does not always coincide to that in space (α) since the mass point is moving in space (α). The value of mean fluctuation velocity (V) is 0 in space (β), but in space (α) the mean fluctuation velocity in x-direction is affected by moving velocity (v) and is not 0. The convolution of the velocity (v) and a function (f) is regarded as the mean fluctuation velocity (fv). The square of inconstancy $(\tau')^2$ is the mean of variation $(V-fv)^2$ which is the difference of mean $(V)^2$ and $(fv)^2$ on the condition that approximate fv is close to τ. The mean of $(V)^2$ is τ^2.

$$(\tau')^2 \fallingdotseq \tau^2 - (fv)^2$$

When the coordinates system (β) stays still to the coordinates system (α), $v = 0$, then (τ') coincides to τ. As v increases, $(\tau')^2$ decreases. $(\tau')^2$ is 0 or more. Hence as equation (48), $f^2 = \tau^2/C^2$ is obtained.

(66) $\qquad \tau' = \tau/[1-(v/C)^2]^{-1/2}$

Therefore, $\tau' = \tau/\gamma$. Ratio τ/τ' is equal to γ function. The inconstancy of mass point (A) observed from the coordinates system (α) is (τ') which is smaller than the inconstancy observed from the coordinates system (β) which is (τ). When $v = 0$, $\tau = \tau'$. Measurement scale is communal to each space. The dimension of inconstancy is velocity, therefore, the velocity observed on the coordinates system (α) becomes slower. "Velocity in moving space retards."

3. Relativity of length

The same discussion as the above is applicable to the insecurity (σ). A mass point (A) stays still at a point of a coordinates system (β) which is moving in the coordinates system (α) in direction x with velocity v. The insecurity probability distribution (Φ) on coordinates system (α) is considered. The insecurity in the coordinates system (α) is not necessarily the same to that in the coordinates system (β). The insecurity (σ') observed from coordinates system (α) has an additional condition that the mass point is moving fixed to the coordinates system (β) where the mean deviation is 0 and the insecurity is (σ). In coordinates system (α) the mass point is in motion with velocity (v), which affects its insecurity. The mean deviation is not 0 and makes small shift in moving direction. Let the mean deviation be the product of the velocity (v) and a function (g). The square of insecurity (σ')^2 in the coordinates system (α) is the mean variation which is mean of $(X - gv)^2$. This mean variation is the difference of square of insecurity (σ^2) in coordinates system (β) and the square of mean deviation which is $(gv)^2$. When gv is close to σ,

$$(\sigma')^2 \fallingdotseq \sigma^2 - (gv)^2$$

When coordinates system (β) stays still to coordinates system (α), (σ') equals to (σ). As (v) increases (σ') decreases. But (σ') is 0 or more, therefore, (v) can have the maximum value which is C.

$$g^2 = \sigma^2/C^2$$

(g) is a constant and following equation (45) is obtained.

(67) $\sigma' = \sigma/[1-(v/C)^2]^{-1/2}$

Hence, $\sigma' = \sigma/\gamma$. Insecurity (σ) has the dimension of length, therefore, the length observed on coordinates system (α) is shorter than that in coordinates system (β). Moving length is shorter than that in standstill. "Length in moving space shrinks."

4. Relativity of time

A space can be defined by one origin point and three basic directions, which is Euclidean space and plural Euclidean spaces can be defined in a Euclidean space. One Euclidean space can move in relative to another origin point. A position is determined in various spaces by the sum of position vector and vector of origin. There is no relativity of time in Galilean relativity in which time is communal throughout to all spaces. The natural time, which is the present instance, is communal to all Euclidean spaces. The time elapse from past to now is considered as negative time and the time duration from now to future is positive time. Then time or time duration from one instance to another instance is measurable continuous physical variable. Past, present and future are considerable at the same time in elapse time or in continuous time. An interval from one instance to another instance in a reference space is a numerical time and is named here conceptual time. Natural time which is measured time has some instability to conceptual time. Time (t) at a point of a coordinate system (α) of which origin is R is considered. The time difference between conceptual and natural time is not always 0. Mean of the difference (T) is 0. The difference probability distributes with a normal function (Υ) centering 0, which is expressed by the following equation. The standard deviation is (v) of which dimension is time. This instability is applicable at any point in space (α).

(68) $\Upsilon = \exp[-T^2/(2v^2)]/[v(2\pi)^{1/2}]$

v is standard deviation of Υ

The equation above is also applicable to any point in coordinates system (β) of which origin is R$'$ when it is stationary to space (α). But when origin R$'$ is in motion in relative to origin R, natural time in space (α) does not change but the instability of natural time in space (β) observed in space (α) increases. T will change. T$'$ is time in space (β) measured in space (α) and its standard deviation is v$'$. When observed from space (α), the mean of T$'$ cannot be 0. The mean depends on velocity. When coordinates system (β) is moving with velocity (v) in direction axis x of space (α), the mean of T$'$ shall be hv, where h is a function. Variance of T and that of T$'$ are the same except the mean value of T$'$. Hence v^2 is the mean of (T$'-hv$)^2. The mean variance of (T$'$)^2 is (v$'$)^2, and the following equation is obtained.

$$v^2 \doteqdot (v')^2 - (hv)^2$$

When $v = 0$, which means space (β) stays still in space (α), v $=$ (v$'$). Instability coincides in both spaces. v^2 has to be 0 or more. Let C be the maximum value of v. Then, h is constant and $h^2 = (v')^2/C^2$, therefore,

$$v^2 = (v')^2 * (1 - v^2/C^2)$$

(69) $v' = v*[1-(v/C)^2]^{-1/2}$

When origin R$'$ is moving in relative to origin R, v$' = $ v$*\gamma$ which means "Time in moving space prolongs." Hence each space has to have its own time-axis.

5. Coordinate transformation

Coordinate transformation in axis x from the coordinates system (α) to coordinates system (β) which is moving parallel to axis x is considered. Let origin R of coordinates system (α) be on origin R$'$ of coordinates system (β) at one instance. After t seconds R$'$ moves on axis x and come to vt. At this instance x' is on x. As the next figure describes, x' equals $(x-vt)$. But x' appears γ times shorter in coordinates system (α).

$$x'/\gamma = x-vt$$

therefore,

(70) $\qquad x' = (x-vt)\gamma$

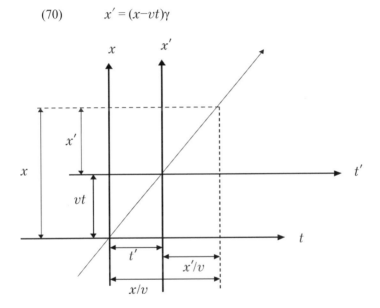

Coordinates transformation to a coordinates system based on the moving origin requires transition of distance, time as well as γ factor. When the transformation is considered in time-axis, the sum of t' and x'/v equals to x/v. But time prolongs in moving coordinates system with factor γ, therefore,

$$(t'+x'/v)\gamma = x/v$$

$$vt' = x/\gamma - x'$$

$$vt' = x/\gamma - (x - vt)\gamma$$

$$vt' = [vt - x(1 - 1/\gamma^2)]\gamma$$

$$vt' = (vt - v^2x/C^2)\gamma$$

$$(71) \qquad t' = (t - vx/C^2)\gamma$$

The time transformation in coordinates system which based on the moving origin require distance, velocity as well as γ function.

M. Biological application of principle of stabilization

1. Stabilization in cell culture

(1) Primary cell number

Cells can proliferate as much as surrounding conditions allow. Cells have intrinsic motivation for dividing and for increasing themselves. But surrounding biological conditions limit the proliferation. The biological conditions provide the aptly balanced proper cell number which is designated here primary cell number. Proper existence of cells depends on the surrounding biological condition of nourishments, space, temperature and so on. When the conditions change, the primary cell number changes. For example, when the nourishment condition deteriorates, cell depletion occurs and the primary cell number decreases. When cell number is less or more than the proper existence level, the cells increase or decrease. The existence level has variation. Cell number which is other than the primary cell number should be regarded as deviation. Current cell number (N_C) is regarded as a deviation (N) of primary existence of cells (N_p). The existence of cells has insecurity (σ) and inconstancy (τ). In consequence, it has continuity (H), variability (Z) and substantiality (Θ).

Applying the existence probability equation,

(72) $\Phi = \exp[-(N_p - N_C)^2/(2\sigma^2)]/[\sigma(2\pi)^{1/2}]$

σ is standard deviation of Φ $\sigma > 0$

Cell number deviation is $N = N_p - N_C$.

(73) $\Phi = \exp[-N^2/(2\sigma^2)]/[\sigma(2\pi)^{1/2}]$

Proliferation rate (R) is the differential of N.

Persistence probability distribution is,

(74) $\Psi = \exp[-R^2/(2\tau^2)]/[\tau(2\pi)^{1/2}]$

(2) Variability

Variability (Z) is obtained which is the quotient of existence probability (Φ) and persistence probability (Ψ) as is shown in equation (7), The both sides of the logarithm of variability equation is taken and differentiated.

(11) $(d\Phi/dt)/\Phi-(d\Psi/dt)-\Psi = (dZ/dt)/Z$

Equations (73) and (74) are substituted into the equation above, then the equation below is obtained.

(75) $(dN/dt)N/\sigma^2-(dR/dt)R/\tau^2 = (dZ/dt)/Z$

Variability (Z) is the proportional constant. $R = dN/dt$. Then, equation below is obtained.

(76) $d^2N/dt^2/\tau^2-N/\sigma^2 = 0$

Let $\tau/\sigma = \omega$, then the dimension of ω is T^{-1}, which is the dimension of angular velocity.

(77) $d^2N/dt^2 = \omega^2N$

Above equation (77) can be changed as follows.

(78) $[d/dt+(\tau/\sigma)][d/dt-(\tau/\sigma)]N = 0$

Right term diverges, therefore, its realistic solution is as follows.

(79) $dN/dt+(\tau/\sigma)N = 0$

The solution of this equation is the following abducting function where N_i is initial value of deviation N.

(80) $\qquad N = N_i*\exp[-(\tau/\sigma)t]$

The amplitude of function (80) converges to 0, which means that the deviation reduces with the attenuation coefficient (τ/σ) and converges to 0.

(3) Continuity

The continuity (H) is the product of existence probability (Φ) and persistence probability (Ψ). The continuity (H) is the proportional constant.

(8) $\qquad \Phi*\Psi = H$

Logarithm of equation (8) is taken and the both sides are differentiated.

$$(d\Phi/dt)/\Phi+(d\Psi/dt)/\Psi = dH/dt/H$$

Equations (73) and (74) are substituted into the equation above and then the equation below is obtained.

$$(dN/dt)N/\sigma^2+(dR/dt)R/\tau^2 = dH/dt/H$$

N is deviation from N_p. $R = dN/dt$ and H is a constant. Hence the equation above becomes the equation below.

(81) $\qquad d^2N/dt^2/\tau^2+N/\sigma^2 = 0$

The ratio of τ/σ is angular velocity (ω). The equation above is an oscillating function. Then the equation below is obtained.

(82) $\qquad d^2N/dt^2 = -\omega^2N$

This solution is a wave function.

(4) Substantiality

The cells increase according to the variability which is equation (79) and stabilize according to the continuity which is equation (81). According to the logarithm of substantiality equation, the sum of these two equations shows the proliferation.

Equation (79) shows the variability which can be changed as follows.

(83) $2\omega dN/dt + 2\omega^2 N = 0$

The continuity contributes to the stabilization. Equation (82) is,

(84) $d^2N/dt^2 + \omega^2 N = 0$

The substantiality of proliferation is the sum of equations (83) and (84), which is the equation below.

(85) $d^2N/dt^2 + 2\omega dN/dt + 3\omega^2 N = 0$

The solution of the equation above is as follows.

$N = \exp(-\omega t) * [N_1 * \cos(2^{1/2} \omega t) + iN_2 * \sin(2^{1/2} \omega t)] + C$ C is a constant

N has to be a real number, therefore, $N_2 = 0$.

When $t = \infty$, N is 0, therefore, $C = 0$.

When $t =$ Initial deviation $N_1 = N_p - N_i$ (initial cell number), Hence

Then equation (86) is obtained which is an abducting wave function.

(86) $N_C = N_p - (N_p - N_i) * \exp(-\omega t)\cos(2^{1/2}\omega t)$

Wave amplitude N reduces and becomes close to 0 and the proliferation is stabilized to cell number N_p.

(5) Overshoot

In cell culture overshooting of cell number is often observed, which means cell number grows over the primary cell number. The cell proliferation accelerated and goes up over the primary cell number and comes back below the primary cell number. It has been often considered that some biochemical process with unknown enzyme makes these phenomena. But the principle of stabilization induces this abducting wave function which explains the overshoot. Cells increase themselves and their growth curves are similar as abducting function. Gompertzian growth model or Logistic growth model shows the similar growth curves as the curves in the variability which equation (80) shows. These growth models do not explain the overshooting. Waving growth in cell proliferation which equation (86) shows has been occasionally observed. Overshooting the primary cell number in the cell proliferation is often observed only at the first wave and the following waves are hardly observed.

2. Tumor growth model with principle of stabilization

(1) Growth volume and vitality

In biological growth analysis it is often considered regenerating cells and tissues take place only with intrinsic motivation. The growth rate is considered constant as long as the tissue property does not change, which is the constant growth rate model. It is simple and good enough for very small term estimations on the tumor growth. Cell division and cell death highly depend on the nutrition supply to cells. Cell death and central necrosis account for the slowing of regeneration rate. Cells have vitality to regenerate themselves. But this vitality depends on not only their own intrinsic motivation but also their intake of such biological substances as nutrition and other stimulating substances. Biological substances from their environment are the source of their vitality. There may be active and passive transport in supplying the biological substances. At the same time cell depletion and cell quiescent occur, which increases as growth volume increases and at the end the intake and output are balanced. Prolific vitality which can be positive or negative is the sum of these cell activity. It has been considered that the vitality decreases as tumors grow, therefore, vitality is inversely proportional to growth volume. And at the same time vitality induces proliferation activity, therefore, vitality is defined to be proportional to growth rate. Hence prolific vitality (P) is designated to the specific growth rate. Growth volume is considered as an accumulation of the prolific vitality. The integration of prolific vitality is designated here growth viability (Q) which is the logarithm of growth volume (G).

(2) Growth volume and growth viability

Viability increases growth volume. Growth volume (G) is dependent to viability (Q). Viability changes growth volume, which is clarified applying variability considering growth volume (G) is dependent factor of viability (Q). dG/dQ is postulated to be the growth persistency (G_q) here. Growth volume (G) has the existence probability distribution with insecurity (σ) and G_q has persistence probability distribution with inconstancy (τ). They have variability (Z).

$$(87) \qquad \Phi = \exp[-G^2/(2\sigma^2)]/[\sigma(2\pi)^{1/2}]$$

σ is standard deviation of Φ $\quad \sigma > 0$

Persistence probability distribution is,

$$(88) \qquad \Psi = \exp[-G_q^2/(2\tau^2)]/[\tau(2\pi)^{1/2}]$$

τ is standard deviation of Ψ $\quad \tau > 0$

Variability is obtained, which is the quotient of existence probability (Φ) and persistence probability (Ψ). It is,

$$Z = \Phi/\Psi$$

The logarithm of the above equation is taken and differentiated with respect to Q,

$$(d\Phi/dQ)/\Phi - (d\Psi/dQ)/\Psi = (dZ/dQ)/Z$$

Equations (87) and (88) are substituted into the equation above, then the equation below is obtained.

$$(89) \qquad (dG/dQ)G/\sigma^2 - (dG_q/dQ)G_q/\tau^2 = (dZ/dQ)/Z$$

Variability (Z) is the proportional constant. And $G_q = dG/dQ$. Then the equation below is obtained.

(90) $d^2G/dQ^2/\tau^2 - G/\sigma^2 = 0$

Above equation (90) can be changed as follows.

$$[d/dQ + (\tau/\sigma)][d/dQ - (\tau/\sigma)]G = 0$$

$dG/dt + (\tau/\sigma)G$ converges to 0 and is not realistic. The realistic solution is as follows.

(91) $dG/dQ - (\tau/\sigma)G = 0$

The solution of this equation is the following equation.

$$G = \exp[(\tau/\sigma)Q]$$

Hence the logarithm of growth volume (G) is as the equation below.

(92) $\log(G) = (\tau/\sigma)Q$

Growth viability is proportional to the logarithm of growth volume.

The above equation is differentiated with respect to time (t) where prolific vitality (P) is dQ/dt.

$$P = dG/dt/G/(\tau/\sigma)$$

Prolific vitality is proportional with growth rate and inversely proportional with growth volume.

Growth viability has the primary value (Q_p) on the balance of intrinsic motivation and its environment. The following equation shows the size of viability deviation (Q_d) from the primary value (Q_p).

$$Q_d = Q - Q_p$$

Growth viability has the existence probability distribution with insecurity (σ).

Prolific vitality (P) has persistence probability distribution with inconstancy (τ). In consequence, it has continuity (H), variability (Z) and substantiality (Θ).

(93) $\quad \Phi = \exp[-(Q-Q_p)^2/(2\sigma^2)]/[\sigma(2\pi)^{1/2}]$

σ is standard deviation of Φ $\quad \sigma > 0$

Deviation (Q_d) is ($Q-Q_p$).

(94) $\quad \Phi = \exp[-Q_d^2/(2\sigma^2)]/[\sigma(2\pi)^{1/2}]$

Prolific vitality (P) is the differential of Q. Persistence probability distribution is,

(95) $\quad \Psi = \exp[-P^2/(2\tau^2)]/[\tau(2\pi)^{1/2}]$

Variability is obtained, which is the quotient of existence probability (Φ) and persistence probability (Ψ) as the following equation shows.

$$Z = \Phi/\Psi$$

Logarithm of both sides of this equation are taken and differentiated.

$$(d\Phi/dt)/\Phi-(d\Psi/dt)/\Psi = (dZ/dt)/Z$$

Equations (94) and (95) are substituted into the equation above, then the equation below is obtained.

$$(dQ_d/dt)Q_d/\sigma^2-(dP/dt)P/\tau^2 = (dZ/dt)/Z$$

Variability (Z) is the proportional constant. $P = dQ/dt$. Then the equation below is obtained.

(96) $\quad d^2Q_d/dt^2/\tau^2-Q_d/\sigma^2 = 0$

Above equation (96) can be changed as follows.

$$[d/dt+(\tau/\sigma)][d/dt-(\tau/\sigma)]Q_d = 0$$

$dQ_d/dt-(\tau/\sigma)Q_d$ diverges and is not realistic. The realistic solution is as follows.

(97) $dQ_d/dt+(\tau/\sigma)Q_d = 0$

The solution of this equation is the following abducting function where Q_i is initial value of deviation Q_d. The ratio of τ/σ is the attenuation coefficient and is replaced with ω here.

(98) $Q_d = Q_i*\exp(-\omega t)$

The deviation reduces with the attenuation coefficient ($\omega = \tau/\sigma$) and converges to 0 and growth viability (Q) becomes Q_p.

    Continuity effects on viability. The continuity (H) is the product of existence probability (Φ) and persistence probability (Ψ), which is expressed with equation below, where continuity (H) is the proportional constant.

$$\Phi*\Psi = H$$

Logarithm of the above equation is taken and is differentiated,

(15) $(d\Phi/dt)/\Phi+(d\Psi/dt)/\Psi = dH/dt/H$

Equations (94) and (95) are substituted into the equation above and then, the equation below is obtained.

$$(dQ_d/dt)Q_d/\sigma^2+(dP/dt)P/\tau^2 = dH/dt/H$$

Q_d is deviation from Q_p. $P = dQ_d/dt$ and H is a constant. Hence the equation above becomes the equation below which is an oscillating function.

$$d^2Q_d/dt^2/\tau^2+Q_d/\sigma^2 = 0$$

The ratio of τ/σ is angular velocity (ω). The equation above is changed as the equation below.

(99) $d^2Q_d/dt^2 = -\omega^2Q_d$

This solution is a wave function.

The cells increase according to the variability which is equation (97) and maintained according to the continuity which is equation (99). According to the substantiality, the sum of these two equations realize the proliferation.

Equation (97) shows the variability which can be changed as follows.

(100) $2\omega dQ_d/dt + 2\omega^2Q_d = 0$

The continuity contributes to the stabilization. Equation (99) is,

(99) $d^2Q_d/dt^2 + \omega^2Q_d = 0$

According to equation (10), the substantiality of proliferation is the sum of equations (100) and (99), which is the equation below.

(101) $d^2Q_d/dt^2 + 2\omega dQ_d/dt + 3\omega^2Q_d = 0$

The solution of the equation above is as follows.

$$Q_d = \exp(-\omega t) * [Q_1 * \cos(2^{1/2}\omega t) + iQ_2 * \sin(2^{1/2}\omega t)]$$

Q_d has to be a real number, therefore, $Q_2 = 0$. When $t = 0$, $Q_1 = Q_i$. Q_i is initial deviation. Then,

$$Q_d = Q_i * \exp(-\omega t) * \cos(2^{1/2}\omega t)$$

Depending on the intrinsic motivations and extrinsic conditions, various final primary viability (Q_p) is expected. Practically Q_p should be obtained applying least squares method. Then, an abducting wave function equation (102) is easier to apply.

(102) $Q = Q_p - Q_i\exp(-\omega t)\cos(2^{1/2}\omega t)$

Wave amplitude $Q_i\exp(-\omega t)$ reduces and becomes close to 0, then the prolif-

eration stabilizes. Viability is proportional to logarithm of growth volume (G).
Then,

$$(103) \qquad \log(G) = \log(G_p) - \log(G_i)\exp(-\omega t)\cos(2^{1/2}\omega t)$$

G_p is primary growth volume, G_i is initial amplitude of growth volume.

3. Human mortality and Gompertzian function

(1) Gompertzian function

In 1825 Benjamin Gompertz published a mathematic function model on
the death rate of population. What he found in the relationship between human
age and the death rate is that exponential of age and logarithm of mortality are in
the linear relationship. The conclusive analysis was stated as follows.

"If the average exhaustions of a man's power to avoid death were such
that at the end of equal infinitely small intervals of time, he lost equal portions of
his remaining power to oppose destruction."

Gompertzian function is usually given as follows.

$$(104) \qquad Y = AB^\wedge c^\wedge x$$

Y is death rate. A is basic mortality which is a constant. c is a constant, x is
age and B is age dependent mortality at age 0. Postulated initial mortality in
Gompertzian function is A*B.

For easier perception logarithm of both sides of the above equation is
shown,

$$(105) \qquad \log Y = \log A + \log B * \exp(x*\log c)$$

As shown in the above equation, logarithm of mortality (Y) and exponential of
age (x) are in the linear relationship. Mortality is increased with age. The phrase
"remaining power to oppose destruction" in the statement above means the vitality to live. The loss of vitality increases infirmity. Infirmity is caused by the de-

crease of vitality and is dependent variable of vitality. Age causes the decrease of vitality which increases infirmity. Infirmity or vitality is the dependent variables of age.

(2) Mortality and infirmity

Infirmity (y) causes existence and persistence probability of mortality. Hence mortality is variated by infirmity. Applying the variability proposition from stabilizing principle mortality is deduced from infirmity. The existence of human mortality has existence probability distribution and insecurity. Y stands for mortality.

$$(106) \qquad \Phi = \exp[-Y^2/(2\sigma^2)]/[\sigma(2\pi)^{1/2}]$$

And the mortality change has persistence probability distribution and inconstancy. F stands for mortality change due to infirmity (y).

$$(107) \qquad \Psi = \exp[-F^2/(2\tau^2)]/[\tau(2\pi)^{1/2}]$$

Variability which is the quotient of existence probability (Φ) and persistence probability (Ψ) is applied and the logarithm of the equation is taken and differentiated,

$$(11) \qquad (d\Phi/dy)/\Phi - (d\Psi/dy)/\Psi = (dZ/dy)/Z$$

Equations (106) and (107) are substituted into the equation above, then the equation below is obtained.

$$(108) \qquad (dY/dy)Y/\sigma^2 - (dF/dy)F/\tau^2 = (dZ/dy)/Z$$

Variability (Z) is the proportional constant. Mortality change $F = dY/dy$. Then the equation below is obtained.

$$(109) \qquad d^2Y/dy^2/\tau^2 - Y/\sigma^2 = 0$$

Above equation (109) can be changed as follows.

$$[d/dy+(\tau/\sigma)][d/dy-(\tau/\sigma)]Y = 0$$

Since Y in equation $[dY/dy+(\tau/\sigma)Y = 0]$ decreases and converges to 0, which is not realistic in human mortality. This is a destructive process, therefore, the following term shall be applied.

(110) $dY/dy-(\tau/\sigma)Y = 0$

The solution of this equation is the following.

(111) $Y = Y_0 * \exp[(\tau/\sigma)y]$

(112) $\log Y = \log Y_0 + (\tau/\sigma)y$

(3) Variability of infirmity

The infirmity is age dependent. Age has strong effect on infirmity. Hence age (x) causes existence and persistence of infirmity.

This equation of infirmity can be deduced applying the variability of the theory of stability. The infirmity has existence probability distribution and insecurity. y stands for the age dependent infirmity.

(113) $\Phi = \exp[-y^2/(2\sigma^2)]/[\sigma(2\pi)^{1/2}]$

And the infirmity alteration is vitality which has persistence probability distribution and inconstancy. V stands for vitality.

(114) $\Psi = \exp[-V^2/(2\tau^2)]/[\tau(2\pi)^{1/2}]$

Variability which is the quotient of existence probability (Φ) and persistence probability (Ψ) is applied and the logarithm of the equation is taken and differentiated,

(11) $(d\Phi/dx)/\Phi-(d\Psi/dx)/\Psi = (dZ/dx)/Z$

Equations (113) and (114) are substituted into the equation above, then the equation below is obtained.

$$(115) \qquad (dy/dx)y/\sigma^2 - (dV/dx)V/\tau^2 = (dZ/dt)/Z$$

Variability (Z) is the proportional constant.

Vitality $V = -dy/dx$ and $dV/dx = -d^2y/dx^2$. Then the equation below is obtained.

$$(116) \qquad d^2y/dx^2/\tau^2 - y/\sigma^2 = 0$$

Above equation (116) can be changed as follows.

$$[d/dx + (\tau/\sigma)][d/dx - (\tau/\sigma)]y = 0$$

Since y in equation $[dy/dx + (\tau/\sigma)y = 0]$ decreases and converges to 0, which is not realistic in infirmity. This is a destructive process, therefore, the following term shall be applied.

$$(117) \qquad dy/dx - (\tau/\sigma)y = 0$$

The solution of this equation is the following.

$$y = y_0 * \exp[(\tau/\sigma)x]$$

This equation is substituted into equation (112) and referred to equation (105).

$$(112) \qquad \log Y = \log Y_0 + (\tau/\sigma)y$$

$$(118) \qquad \log Y = \log Y_0 + (\tau/\sigma) * y_0 * \exp[(\tau/\sigma)x]$$

$\log Y_0$ is a constant which is $\log A$ and $(\tau/\sigma) * y_0$ is initial value of y which is $\log B$.

And τ/σ is $\log c$, hence

$$Y = A * B^{\wedge}\exp(x * \log c)$$

In Gompertzian function the continuity of the human mortality is neglected which may give a waving unstable mortality curve.

4. Gompertzian function growth model

(1) Application of Gompertzian function

Gompertzian function is highly applicable to various phenomena since it is a function based on the proposition of variability from stabilization principle. It has been originally developed in 1825 to estimate human mortality. In 1926 Wright and Sewall made a book review in *Jour. Am. Stat. Assoc.* in which the application of Gompertzian function to such biological developments as cell proliferation, organ development and so on. In the mortality curve the exponent of exponential term is positive. In the biological development curves the exponent is negative. A tumor growth model is obtained here applying Gompertzian function expressed in logarithm.

$$\log(Y) = \log(A) + \exp(x * \log c) \log(B)$$

$\log(Y)$ is replaced with the growth volume G, x with time t, and $\log c$ has to be a negative constant and is replaced with $(-r)$.

At $t = \infty$ Growth volume G has to be the maximum or primary growth volume G_p, therefore, $\log(A) = G_p$. At $t = 0$ G is the initial volume G_i, then $\log(B) = -(G_p - G_i)$.

$G_p - G_i$ is initial deviation D_i. D stand for the volume deviation from the primary volume G_p. Hence the above equation is rewritten as follows.

$$(119) \qquad G = G_p - D_i * \exp(-r * t)$$

The growth volume increases with the attenuation coefficient $(-r)$ and converges to G_p.

The same equation is obtained when G is applied to the variability as follows.

Gompertzian function is essentially for the function of destructive variability. Tumor growth is a constructive process where the coefficient of variability equation has to have the opposites of those in destructive process.

(2) Variability in Gompertzian function

Growth volume is the difference of primary volume and deviation volume. Since primary volume is a constant, the variability of growth volume is the same to that of deviation volume. For convenience deviation volume is employed here. The existence of deviation volume D has existence probability distribution with insecurity (σ).

$$(120) \qquad \Phi = \exp[-D^2/(2\sigma^2)]/[\sigma(2\pi)^{1/2}]$$

And the deviation change rate R has persistence probability distribution with inconstancy (τ).

$$(121) \qquad \Psi = \exp[-R^2/(2\tau^2)]/[\tau(2\pi)^{1/2}]$$

Variability (Z) which is the quotient of existence probability (Φ) and persistence probability (Ψ) is employed here.

$$(\Phi/\Psi = Z)$$

And the logarithm of this equation is taken and differentiated.

$$(11) \qquad (d\Phi/dt)/\Phi - (d\Psi/dt)/\Psi = (dZ/dt)/Z$$

Equations (120) and (121) are substituted into the equation above, then the equation below is obtained.

$$(122) \qquad (dD/dt)D/\sigma^2 - (dR/dt)R/\tau^2 = (dZ/dt)/Z$$

Variability (Z) is the proportional constant. Deviation change rate R is $R = dD/dt$. Then the equation below is obtained.

$$(123) \qquad d^2D/dt^2/\tau^2 - D/\sigma^2 = 0$$

Above equation (123) can be changed as follows.

$$[d/dt+(\tau/\sigma)][d/dt-(\tau/\sigma)]D = 0$$

Since D in equation $[dD/dt-(\tau/\sigma)D = 0]$ increases and diverges, which is not realistic in growth, therefore, the following equation is applied.

(124) $dD/dt+(\tau/\sigma)D = 0$

The solution of this equation is as follows.

$$D = D_i*\exp[(-\tau/\sigma)t]$$

D_i is the initial deviation volume which is (G_p-G_i). G_i is initial growth volume. G is an increasing function and D is deviation volume from G_p, therefore, G is,

(125) $G = G_p-D_i*\exp[(-\tau/\sigma)t]$

This equation is the same as the one derived from Gompertzian equation where $\tau/\sigma = r$ $(r = -R)$.

(3) Continuity

Growth volume is also affected by continuity. The continuity (H) is the product of existence probability (Φ) and persistence probability (Ψ). The continuity (H) is the proportional constant.

(8) $\Phi*\Psi = H$

Logarithm of the above equation is taken and the both sides are differentiated,

$$(d\Phi/dt)/\Phi+(d\Psi/dt)/\Psi = dH/dt/H$$

Equations (120) and (121) are substituted into the equation above and then the equation below is obtained.

(126) $(dD/dt)D/\sigma^2+(dR/dt)R/\tau^2 = dH/dt/H$

$R = dD/dt$ and H is a constant. Hence the equation above becomes the equation below.

(127) $d^2D/dt^2/\tau^2 + D/\sigma^2 = 0$

The ratio of τ/σ is angular velocity (ω). The equation above is an oscillating function. Then the equation below is obtained.

(128) $d^2D/dt^2 = -\omega^2D$

This solution is a wave function.

(4) Substantiality

      According to the substantiality, the sum of these two equations shows the proliferation.

      Equation (124) shows the variability which can be changed as follows.

(129) $2\omega dD/dt + 2\omega^2D = 0$

The continuity contributes to the stabilization.

(128) $d^2D/dt^2 + \omega^2D = 0$

According to equation (10), the substantiality of proliferation is the sum of equations (129) and (128), which is the equation below.

(130) $d^2D/dt^2 + 2\omega dD/dt + 3\omega^2D = 0$

The solution of the equation above is as follows.

$$D = \exp(-\omega t)*[C_1*\cos(2^{1/2}\omega t) + iC_2*\sin(2^{1/2}\omega t)] + C$$

C_1, C_2, C are constants

D has to be a real number, therefore, $C_2 = 0$.
When $t = 0$, $C_1 = D_i$. D_i is initial deviation. Then,

$$D = D_i * \exp(-\omega t) * \cos(2^{1/2}\omega t)$$

$$D_i = G_p - G_i$$

Then equation (131) is obtained, which is an abducting wave function.

(131) $$G = G_p - (G_p - G_i)\exp(-\omega t)\cos(2^{1/2}\omega t)$$

Wave amplitude D reduces and becomes close to 0, then the proliferation is stabilized at primary volume G_p. In practice G_p should be obtained applying the least squares method in which the waving phenomenon and overshooting process in proliferation can be neglected.

Epilogue

Man has had an ancient conception that God created the whole world and divine volition controls the whole or majority of natural events. It has been thesis that divine creation has no instability and is perfect. But the intelligence of people has investigated to find several natural laws, which has been antithesis. The whole or majority of nature is created by the contingency. The contingency brings forth various laws of nature which are always sought for to describe the nature. As synthesis the laws from contingency could be regarded as divine volition. The essentiality of events is stabilization which can describe fundamental laws of nature. At the beginning of freshman of medical school author looked in a book in a student library. The book explained on Lorenz function which was unbelievable and annoyed author for a long time. The answer was contingency and stabilization. Contingency which can be expressed with normal function brings forth stabilization. Existence and persistence in stabilization principle which has been stated in prologue were realized in the processing a set of oncology data. Some fifty years ago author was trying to develop numerical formulas of tumor growth models processing clinical data and realized fluctuations of growth curves which used to be considered as measurement errors. They appeared waving and the growth curves appeared to contain wave function, which easily explain the overshooting of the growth curves. Repeating statistical processing the data, it is realized that the waving is explained when the stability of a mass is expressed with a normal distribution function which explained variability and continuity in the tumor development. Author also realized this concept explains the presence of the universal gravity law. Axiom of insecurity deduces the universal gravity and axiom of inconstancy deduces the special relativity. Variability and continuity of events are realized here, which substantialize the existence of events. Variability and continuity are formed from the above two axioms.

Physical existences have substantiality. Substantiality exists even with no mass. Some mesons and quarks exist without mass. Space and its parts exist physically, therefore, they have substantiality and various physical activities. Events of low continuity change easily with high variability but stay long if the substantiality changes into low variability. They say,

> "Fragile things change easily but they stay long with improved conditions."

Events with low continuity will change into high continuity reducing variability with favorable conditions to exist easier. Analyzing concept of variability and continuity, the substantiality and stabilization were realized. Summarizing these propositions of variability and continuity induced stabilization principle which is essential in actual existence of events. It is expected that this stabilization principle and propositions stated above are accepted and will be of use in various studies.

安定化原理

プロローグ

　人の考える存在即ち主観的存在はスカラー量を用いて正確に認識される概念上の存在で、自然の客観的存在は人間の思考とは独立の存在であり、偶然性が介在する自然の法則に従う。主観的存在は安定であり自然の存在とは質的な差異がある。自然の存在とその変化は偶然であり、その偶然性のもたらす安定化、即ち事象の確率の上昇に基づく法則による。位置と運動はスカラー量としてではなく統計量として扱われるべきである。従来は質量の有る存在について検討したのであるが、質量の無い自然の存在についても検討するべきであり、また自然のもたらすその他の現象についても検討する。自然は偶然がもたらす安定化の産物であり、変動は安定化、即ち存在確率、持続確率の上昇の原則に基づいている。自然のもたらす事象の本態を統計量とし安定化原理に基づき解析した。この安定化原理の本態は次の散在性と揺動性の二つの公理から成る。

　　1　散在性：事象は存在する。存在には統計的な散在性がある。
　　2　揺動性：事象は持続する。持続には統計的な揺動性がある。

　事象は状況が変化すれば変動して安定化し存続する。事象には変動性、存続性、実在性の公理がある。変動とは安定化することであり存在の確率と持続の確率は比例する。安定な事象の存続には存在の確率と持続の確率が反比例する。存続性の低い事象は存続し易い状況に向かって変動性を減少させ存続性を高める。事象の実在においてはこれら変動性と存続性は質量の有る無しにかかわらず反比例して実在性をもたらす。

　上述の概念、即ち安定化原理に基づく理論は諸般の基本的な法則等に矛盾していないことを示している。万有引力の法則は散在性要素に起因

する。特殊相対性理論の成因も揺動性要素に在る。これら二つの公理か
ら変動と存続の命題が導かれ事象の変化と存在を説明する。事象は変動
して安定化し存続する。変動性と存続性が反比例しているのが事象の実
在性である。この存在実態は質量が有る事を前提としない。質量の無い
空間の存続性はエネルギーの発生に関与している。偶然性に基づく安定
化原理は従来の基本法則を説明することができる。安定化原理とは事象
には存在の確率と持続の確率が有ることで、これに基づく命題は事象が
変動して存続する、実在の事象には変動性と存続性が共存する。変動性
が色々な事象に応用され各種研究にも利用される事を期待する。変動性
と存続性の概念は認識されるべきであり、多くの項目に応用され得ると
考える。この統計的思考に基づく事象の安定化の概念は細胞増殖や腫瘍
増殖を数式モデル化する研究において認識されたが基本的な物理法則も
説明し、他の科学的事象の説明に応用できるであろう。

目 次

A. 存在の本質

１. 空間と存在の概念

⑴ 空間

　空間が如何なる存在であるかは思考上の把握である。一次元から多次元まで考え得る。高次元空間は自然現象や社会現象の解析に用いられている。これらは概念としての空間である。概念空間内の点は正確な位置や運動に完全な安定性が有る。位置的時間的に安定であり空間には時間軸も伴う。

　宇宙の大空間が如何なる存在であるかは把握困難である。宇宙の大空間がどんな存在であるとしても、この近傍の空間は三次元空間であると考えられる。物の存在を通常ユークリット空間と呼ばれている三次元空間内で主観的にも客観的にも考える。点の集合である空間は原点一つと三つの基本方向で定められる。この原点の位置と運動は主観的概念空間では完全に安定である。現実の物理空間では客観的に認識されるべきである。宇宙の大空間が如何なる存在であるかは把握困難であるが、ここ近傍の空間は客観的三次元空間であり主観的概念空間に対して位置的時間的不安定性を伴う実在性が有る。よって現実の空間は実在空間であり、その実在性に起因する熱の空間である。実在空間ではその位置と運動は概念空間に参照して不安定性が有るゆえ実在空間の全ての点に散在性と存在確率、揺動性と継続確率の不安定性が有り、実在空間の全ての点は実在性である。

　概念空間でも実在空間でもその中に二次的空間があり得る。二次空間の参照枠は一次空間である。一次空間の参照枠は概念空間であるから実在空間の位置と運動には不安定性があり、その結果、実在性が存在するのである。

　現実の物理空間即ち実在空間は常に三次元空間であり固有の時間を伴う。

　物の存在は常に三次元空間とその時間で考えられるので時間の概念が必要となる。空間で考えられる時間は経過時間である。経過時間はここでは時間と呼ぶが、ある瞬間から他の瞬間までの間隔である。実在空間には常にその固有の時間が有る。一つの一次空間には数個の二次空間があり得る。それぞれの実在空間は固有の時間が有り、その値は統計量であるが概念空間では時間は全ての空間で共通でスカラー量である。それゆえ実在空間での動態事象では時間の相対性がある。時間は一空間内での独立変数である。例えば、一基本方向の距離が時間の従属変数と成り得る。空間と時間が存在の枠組みを構成する。空間と時間が物理的事象の存在を表すのに必要とされる。質点の存在は、その質点の属する実在空間の空間時間枠で表される。

⑵　存在

　ユークリット空間内の一点には位置と運動があると考える。一点の位置には他の点との相対性が有り、運動はその相対性の変化である。一点の位置が変化すれば、例えば原点との相対性が変化する。この空間や位置、運動が正確で安定とする認識が概念的空間における存在の概念である。現実空間内の点を認識する為の概念でもあるが、そこにはある程度の不確実性がある。

　三次元空間内に一形体が存在している場合、実在形体上のある一点はこの空間における概念形体上の対応する点と完全な精度で不確実性なく一致するとは言えない。一点が概念上でなく実在の一点で参照枠内の対応する点があり得るが必ずしも一致しない、ずれもあり得る。この場合の参照枠が実在空間であっても一致とは概念であり完全なものではない。一致にはずれがあり得る。一致とは概念で完全な一致ではなく偏位を伴う。その精度は統計的である。偏位で統計的な精度が低下する。そ

れが存在の散在性である。期待位置は存在確度分布の中心であり、そこが高確度である。存在確度の差があると求心力が発生する。これが引力の源である。実在空間においては空間が存在を規定し時間が変化を規定する。物理的な物体は実在空間を参照する位置と運動があるからずれや揺らぎが起こる。またそれらにより散在性や揺動性が生じる。存在は時間により変化する。空間が散在性をもたらし時間が揺動性をもたらす。

　概念上形体には点や線もある。線は点の連なりであり、点や線や立体形体は点または点の集合体と考えられるゆえ、三次元空間内で表すことができる。個体は固形のエネルギー即ち物質であるが実在点の集合であり、その存在を参照枠内で表し得る。実在空間内の物体は人間の概念空間内の存在と一致するとは限らない。人間の思考は計測器と同じで量はスカラー量である。この計測器で測られる実在量は統計量である。この意味で散在性を伴う点の集合体は物理的存在、即ち物体である。実在の物体は位置と運動があるが一定の運動を継続するには相対的な困難さがある。それが質量である。質量は大きさや種類の違いによる運動速度の変化の困難さの比較のことである。本質的には持続確率分布がこの困難さの違いを表している。

　物理的な物体はしばしば、その形や大きさを無視して幾何学的な一点で代表される。幾何学的一点が物体の質量を代表する。それが質点である。エネルギーの存在に関しても幾何学的な一点に代表させ、物理的なエネルギー量もその形態や大きさも認識しがたいものであるがゆえに、この一点に与えている。

２．安定化の要素

　事象は存在して変化する。事象には一次量と二次量があり、一つの共通要素が双方に不確定性をもたらしている。例えば電気量について、電気量は一次量で電圧に規定されるが変化もする。この電気量の二次量は

電流であり電圧に規定される。共通要素は電圧である。二次量は一次量の変化である。質点もその例である。質点はある位置にある。これが一次量であり距離に規定される。この位置から変動する可能性もある。それが運動である。この運動が二次量であり距離にも規定される。距離が共通の要素であり、質点の存在と運動の共通要素となっている。質点は常にあるべき点上に正確にあると考えるのは概念であり、現実はそのあるべき点からの偏位があり得る。この偏位量は統計量である。存在確度は偏位量による。

存在精度は完全ではない。偏位が小さい存在確度は高いが完全ではない。大きな偏位の確度は低い。少偏位の確率は比較的高く存在確率分布は偏倚なしを中心とした正規分布となる。

すべての事象は変化か持続する。変化は二次量である。存在の変化も統計量である。急激な変化の確率は低いであろう。緩やかな変化の方が比較的に確率は高いであろう。変化のない状態は持続である。その確率は高いが100％ではない。持続の確率分布は変化なしを中心とした正規分布となる。

事象には二つの重要な要素がある。それらは存在と持続である。質点を例にして考えれば、質点には位置と運動がある。運動とは位置の持続性である。それは時間因子で距離を微分して得られる。位置には存在確率があり、運動には持続確率がある。存在には散在性があり、持続には揺動性がある。この二要素量は統計量である。事象には存在と持続の二つの重要要素があり、それらは散在性と揺動性の統計量である。

これらは安定化の過程であり次の二つの公理からなる安定化原理により明確に説明される。

1）散在性：事象は存在する。存在には統計的な散在性がある。
2）揺動性：事象は持続する。持続には統計的な揺動性がある。

3. 存在の散在性と揺動性

　人間の考える位置即ち点は主観的存在で概念上の一点で時空内の一点に概念上正確に位置を取ることができ、また、その点は時空内を概念上自由に動き得る。完全な精度で、不確実性はなく、時空内の一点上に存在し得る。また、運動にも完全な安定性がある。自然に基づいた客観的実在の点は質点で最も単純な存在と考えられる。

　質点は人間の自由な思考通りには存在し得ないし動き得ない。人間の思考とは独立の存在であり、自然の法則に従う。概念上の存在は人為的であるが、自然の法則に従う存在には偶然性が介在する。物体上の一点は人間の思考に対して独立で物体固有の散在性を有している。物体上の一点は物理的空間内の一点上に存在するとしても、その存在には、概念上の一点の存在とは質的な差異があり、ずれや揺らぎが存在する。質点の存在は位置と運動を有するが、存在位置にはずれやばらつきが起こり得る。これを散在性と呼ぶ。また、持続する運動にも不安定な揺らぎが起こり得る。これを揺動性と呼ぶ。

B. 存在の散在性

⑴ ばらつき

　まずは比喩を用いて感覚的に存在の概念を考えてみる。ある平面上に
ゴルフボールを置くとする。その平面の座標を考え、そのボールの重
心を平面の座標（0, 0）の位置に置いたと考えることができる。この存
在は思考上点（0, 0）の位置に存在すると考えられるとしても、現実に
ボールを100回置いて、皆同じ位置に置けると考えるのは概念上可能で
あっても現実にはない。現実は必ず誤差を伴う。今、大学生がこの行為
を行ったとしたら、ボールの位置の最尤値は点（0, 0）であるとしても、
ある程度のばらつきを生じる。同じ行為を小学生が行えば、ボールの位
置の最尤値はやはり点（0, 0）であろうが、もっと散在している。ばら
つきは大きくなるであろう。ゴルフボールの重心の存在が点（0, 0）の
位置であっても質的に違う存在である。いずれもボールの重心の位置が
点（0, 0）で代表される存在である。中心は点（0, 0）に存在しても、あ
る標準偏差値をもって、その周りに存在確度を分散している。前者のほ
うがより精確で小さな近傍内に集在している存在であり、分散値が小さ
い。重心点の最尤値（0, 0）の近傍内への出現頻度が高いので存在確度
が高く高精度な存在と表現する。後者の方が誤差の大きな精度の低い散
在性の高い存在である。存在の精確性に差異があり、後者の標準偏差は
大きくばらついているので、上記重心点は近傍内への出現頻度が比較的
低い。

⑵ 一点上の存在

　逆の方向から考えて、ゴルフボールの重心が今点（0, 0）の上に在る
とする。この重心がゴルフボールの存在を代表できると考える。ゴルフ
ボールの重心を質点と考えて、この質点が空間内の一点（0, 0）に存在

するとする。この重心点が点（0, 0）の上に存在していると考えるのは概念であって、現実に点（0, 0）上に在る確率はかなり高いが散在性もあり、ずれの可能性を否定できない。その存在位置に不確実性がある。誤差即ち不確実性があり存在確度を伴った状態である。存在状態に精度を伴う。その質点の散在の確度分布では存在が期待される点（0, 0）上での確度が極大で、その点を中心に周りにより低い確度のばらつきがあることである。これを存在の散在性と呼ぶ。その精度に高い状態や低い状態があり得る。

　素粒子の位置に関してはそのように考えやすい。電子やほかの素粒子のある瞬間での位置を考える場合、その位置を計算することは出来るが、全ての粒子が計算された位置にあるわけではない。その計算値にはばらつきを伴う。それが散在性である。計算された位置は最尤値であり、その周りに粒子は散在していて統計的な存在である。

⑶ 存在確度分布

　存在の確度分布とは本来位置またはその偏位への存在の可能性で、偏位を確率変数とする確度分布である。本来の存在位置の確度が最尤値であり、これを中心に正規分布を成す。その最尤値における存在の確度が存在確度であり存在の精確さである。これが高いほど安定な存在と言える。この確度分布の標準偏差値を散在性と呼び、これが小さいほど最尤値における存在の精度が高い状態で、質点はより安定に存在する。この標準偏差値の逆数を精度と定義する。煩雑さを避けるために、その質点（例えばゴルフボールの重心点）が存在する点の x 軸方向のみの状態確率について考える。この質点の生起率は x 軸上どこでも均一であるので、その確度分布（Φ）は、ある標準偏差値を持って、期待値（a）を中心に正規分布をしている。

$$(1) \qquad \Phi = \exp\left[-(x-a)^2/(2\sigma^2)\right]/\left[\sigma(2\pi)^{1/2}\right]$$

σはΦの標準偏差値

X を点 x とその質点の存在期待値（a）からの偏位とすると、X = $x-a$ だから、

$$(2) \qquad \Phi = \exp\left[-X^2/(2\sigma^2)\right]/\left[\sigma(2\pi)^{1/2}\right]$$

存在位置即ち期待値上での存在の確度は、この散在確度分布の X = 0 の時で、$1/\left[\sigma(2\pi)^{1/2}\right]$ である。この値が高いほど少しのずれで確度は急に低下するので、ずれの存在しない期待値（a）上への存在する精確な状態と言える。この標準偏差値（σ）が大きいほど存在の確度は下がるゆえ、（σ）は散在性を表し、その逆数（$1/\sigma$）は存在の精度と呼ぶ。

⑷ 生起率と散在性

質点のエネルギー量を E とし、単位エネルギー当たりの生起率を P_C とすると、標準偏差値（σ）と生起率（P_C）の関係は、

$$(3) \qquad \sigma^2 = P_C(1-P_C)/E$$

であり、P_C は 1 に比して充分小さい値であるから、

$$(4) \qquad \sigma^2 \fallingdotseq P_C/E$$

生起率を総エネルギーで除したものの平方根が標準偏差値となる。よって、単位エネルギーの存在確度分布における平均分散はおよそ生起率に等しいゆえ、生起率の均一な空間内の一点に存在する質点は、より高い存在確度の位置はないので安定であり、他の位置に移動を起こすことはない。たとえ起こしたとしても均一な空間であり、その散在性には変化がなく同じ中心力が作用する。質点に作用する求心力はどの方向からも均等であり、その位置に安定に存在している。この存在点上での存在確度が極大である。その存在確度は100％ではなく位置移動を起こす可能

性も共存している。空間内の生起率が不均一で、質点の存在する位置
(b) より低い生起率、即ち、より高い存在確度の位置 (a) が発生する
とすると、その位置はより高い存在頻度の位置なので質点は移動を起
こす。もとの位置 (b) はこの位置 (a) の存在確度分布に基づく偏位
(X) となり、求心力により移動して安定化する。安定後の質点は存在
確度が上がるので安定性はより高くなる。

⑸ 存在の散在性

　事象は存在する。存在には統計的な不安定性である散在性がある。存
在には存在の確率分布がある。存在は統計量であり統計学的な不安定性
が内在する。それが散在性 (σ) である。存在確率分布は正規分布であ
り、偏位 (X) への存在確度 (Φ) は、

$$(2) \qquad \Phi = \exp[-X^2/(2\sigma^2)]/[\sigma(2\pi)^{1/2}]$$

期待値上への存在精度が完全なことはなく散在性が常に伴っている。散
在性は標準偏差値 (σ) で表す。

C. 存在の揺動性

(1) 持続性

　同じ精度の存在としても変化し易い存在と持続する存在とがある。揺らぎを感覚的に考えてみる。ゴルフボールの代わりにピンポンボールを位置 $(0, 0)$ に置く行為を行ったとする。大学生が上手に位置 $(0, 0)$ に100回置いて、ゴルフボールの時と同じ高精度な存在を得られたとしても、ゴルフボールの場合より揺らぎ易い。置いたときの位置移動が起こり易い。即ち定着性が悪い。ゴルフボールを置いた時の方が揺らぎが少なく静止し易い。静止の弾みが大きく、この位置での揺らぎが0である確率が高い。同じ精度の位置 $(0, 0)$ 上への存在であっても存在の定着性に差異がある。今ゴルフボールの重心が点 $(0, 0)$ 上に在り静止している存在であるとすると、その実態は揺らぎの速度が0、即ち固定である確度が最も高い存在である。固定の確度の高いほど静止の弾みが大きい存在である。次にゴルフボールの代わりにピンポンボールがあり、その重心が点 $(0, 0)$ の上に在るとする。ゴルフボールに比して揺らぎの可能性は高く、揺らぎのない固定状態であるとしても静止の弾みは比較的小さく、存在位置への定着が弱い。ゴルフボールの方が落ち着きがあり、揺らぎの可能性の少ない存在と言える。

(2) 揺らぎ

　存在には期待している点上にあるかどうかという位置的な不安定性即ち散在性だけでなく、その点での揺らぎを起こす可能性の質的な不安定性も内在する。これを存在の揺動性と呼ぶ。揺らぎは持続、即ち揺動が0の固定速度を中心に正規分布を成す。その標準偏差値で揺動性を表す。その標準偏差値の逆数を強度と定義する。強度が低く揺らぎ易い存在や、強度が高く揺らぎ難い存在がある。いわゆる重いボールは強度が

高い存在で、静止の弾みが比較的大きく揺動を起こし難く持続する。ピンポンボールは静止の弾みが小さく揺動を起こし易い存在で、揺動が０の固定確度が低くなる。この相違は存在の強度の違いと表現することができる。強度が高いとは質点と位置の結びつき（静止の弾み）が強い状態と言える。点の集合としての形態の変化についても同様である。一つの形態の代表的な複数の点の相互関係が変化し難い状態が強度の高い存在である。

　一点上の存在を考えると、ずれの可能性だけでなく、揺らぎの可能性も在る。一般的な存在には精度と強度の二つの安定要素がある。

(3) 持続確度分布

　質点には位置と運動の二要素がある。運動の不確実性は揺動なしの点を中心に正規分布を成している。その標準偏差値が揺動の程度を示す。この持続の確度分布とは揺らぎの可能性で、存在の確率変数（位置）の時間変化率、即ち運動を確率変数とする。この運動の不安定性をここでは揺動速度または揺動と呼ぶ。静止は運動（０）が最尤値であり、その確度が静止確率である。持続確率は最尤値の周囲に正規分布を成す。その標準偏差値が小さいほど持続の可能性が高く、堅固な存在と成る。それゆえ、この標準偏差値の逆数を強度とする。ユーグリッド空間内の一点（a）に静止している場合はその運動速度は０である。静止状態といえども揺動の可能性はある。均一の空間の一点に質点が静止している場合質点周囲への揺動はどの方向にも均一である。その結果その変動は０である。x軸上の点（a）の位置に存在する質点の偏位を X とすると、揺動（V）は次の式で表される。

$$(5) \qquad V = dX/dt$$

この質点の揺動は如何なる方向にも速度０を中心とした正規分布を成す。この確度分布（Ψ）が持続の確度分布である。この確度分布のば

らつきの幅、（標準偏差 τ）が揺動性であり、物体の強度（持続の弾み）に反比例する。x 軸上で考えると、$x = a$ に静止している質点の揺動はある標準偏差値をもって揺動 0 を中心に正規分布を成すので、

$$(6) \qquad \Psi = \exp[-V^2/(2\tau^2)]/[\tau(2\pi)^{1/2}]$$

$V = 0$ のときの確度は $1/[\tau(2\pi)^{1/2}]$ で、これが高いほど少しの揺らぎでその確度は急速に低下する。

⑷ 存在の揺動性

　事象は持続する。持続には統計的な不安定性である揺動性がある。持続には揺動の確率分布がある。

　事象は運動を持続する。その持続は統計量であり統計学的な不安定性が内在する。それが揺動性である。持続確率分布は正規分布であり、揺動（V）での揺動確度（Ψ）は、

$$\Psi = \exp[-V^2/(2\tau^2)]/[\tau(2\pi)^{1/2}]$$

であり、期待値上への存在強度は完全なことはなく揺動性が常に伴っている。揺動性は標準偏差値（τ）で表す。

D. 質点の実態

⑴ 質点の変動性

　概念上の形態は概念空間内を自由に動き得る存在であるが、物体を構成する形態が実在空間内を自由に動き得るわけではない。存在確度の相違があるからである。運動の実態は基準点と形態上の各点との相対関係の時間的な変化であり、存在確度の変化を伴って変位を起こす事象の変動である。運動には時間を伴う。運動を考えるにあたっては時間の概念が必要となる。ある時刻から他の時刻への経過が時間であり、一般的であるガリレオ相対論では如何なる空間でも共通であると考えられている。変動は存在確度分布の偏位による運動であり、その結果として存在確度の変化も生じて存続する。散在性が高い事象は変化し易いが揺動性が高ければ、なお変動し易い。これが変動性である。変動性は存在確度と持続確度の比である、と定義する。

$$(7) \qquad Z = \Phi/\Psi$$

⑵ 質点の存続性

　安定な存在が実在であり変動性と存続性から成る。変動の本質は安定化である。散在性が高い事象でも、揺動性が低ければ存続し易いのが存続性である。事象は変動するかまたは存続する。事象は状況が変われば変動して存続する。事象が存続するには散在性が高い時は揺動性が低くなければならない。揺動性が高い時は散在性が低くなければならない。存続性においては散在性と揺動性は逆方向である。よって、存続性では散在性と揺動性は反比例している。存続性は存在確度と持続確度の積である、と定義する。要約すると、事象はその散在性と揺動性に比例して変動し、散在性と揺動性が反比例して安定に存続する。

$$(8) \qquad H = \Phi * \Psi$$

⑶ 質点の実在性

　事象の実態は一般にスカラー値で把握される。しかし、それらの値は変動性と存続性のバランスで定まる統計的確率が最も高い値と把握するべきである。事象の実質は変動性と存続性とから成るが、変化しつつある事象の実在性は大きな変動性と小さな存続性から成る。また、安定した事象の実在性は大きな存続性と小さな変動性から成る。変動性と存続性は逆方向である。変動性が高い時は存続性が低く、変動性が低い時は存続性が高いのが事象の実在性である。変動性と存続性は反比例していて変動性と存続性の積が事象の実在性である。

　物の安定性について次のことが言える。

　　「壊れ易い物は存在し難いが、適切な条件下では存在する。」

　この意味は、存続性の低い事象は変動するが変動性の低い状況を得れば存続するということである。実在性（Θ）を変動性と存続性の積と定義する。存在状況が変化すると、事象は変化して、一般に存続し易い安定な存在状態に変動する。変動性が減少して存続性が上昇し安定化する。この経験則は物理現象のみでなく多くの自然現象に適応できるであろう。また、人間の社会現象や精神現象等にも適応できるであろうが、ここでは質点の存在と運動に関して考える。実在性（Θ）は存在の特性であり定数である。実在性（Θ）では変動性と存続性が反比例しているゆえ、次の式で表される。

$$(9) \qquad Z * H = \Theta$$

両辺の対数を取る。

$$\mathrm{Log}(Z) + \log(H) = \log(\Theta)$$

この両辺を微分すると、Θ は定数であるから $d\Theta/dt/\Theta$ は 0 。

(10)　　　$dZ/dt/Z + dH/dt/H = 0$

状況に変化のない時は存続性（H）は変化しない。よって、$dH/dt/H$ は 0 。ゆえにこの場合、$dZ/dt/Z$ も 0 。状況に変化のない場合は H も Z も定数である。

E．回転と安定化

1．事象の回転

⑴ 変動性と回転

存在確度と持続確度は変動性においては比例する。下記の式の如く変動性（Z）は比例定数である。

$$\text{(7)} \qquad \Phi/\Psi = Z$$

両辺の対数を取り微分すると下記の式となる。

$$\text{(11)} \qquad (d\Phi/dt)/\Phi - (d\Psi/dt)/\Psi = (dZ/dt)/Z$$

散在性変化率と持続性変化率の差が変動性の変化率となる。式（2）と式（6）を導入すると下式を得る。

$$\text{(12)} \qquad (dX/dt)\,X/\sigma^2 - (dV/dt)\,V/\tau^2 = (dZ/dt)/Z$$

X は x 軸上の点 a からの偏位であり V = dX/dt である。Z は定数であるから下式を得る。

$$\text{(13)} \qquad d^2X/dt^2/\tau^2 - X/\sigma^2 = 0$$

τ/σ = ω とすると、ω の次元は T^{-1} で角速度の次元となり、下式を得る。

$$\text{(14)} \qquad d^2X/dt^2 = \omega^2 X$$

点 x = a に存在する質点は回転していて正の加速度、即ち遠心力を起こしている。

⑵ 存続性と回転

存続性とは安定した存在である。存続性（H）は存在確率（Φ）と持続確率（Ψ）との積である。存続性（H）はその比例定数である。質点（A）が x 軸上の点（a）に存在する時の存続性は下式で表される。

$$(8) \qquad \Phi * \Psi = H$$

上式の対数を取り微分すると下式を得る。

$$(15) \qquad (d\Phi/dt)/\Phi + (d\Psi/dt)/\Psi = dH/dt/H$$

式（2）と式（6）を上式に代入すると下式を得る。

$$(16) \qquad (dX/dt)X/\sigma^2 + (dV/dt)V/\tau^2 = dH/dt/H$$

X は x 軸上の点（a）からの偏位で、揺動 V は V = dX/dt で、存続性（H）は定数であるゆえ下式を得る。

$$(17) \qquad d^2X/dt^2/\tau^2 + X/\sigma^2 = 0$$

揺動性と散在性の比（τ/σ）は角速度（ω）であるから、上式は波動関数である。質点は点（a）上を回転している。下式を得る。

$$(18) \qquad d^2X/dt^2 = -\omega^2X$$

存続性にも波動性がある。点（a）上を回転している質点には負の加速度があり求心力が起こる。安定した質点には角速度があり点（a）上で回転している。式（14）は点（a）上で回転している質点には正の加速度があり遠心力を示している。そして式（18）はその質点には負の加速度もあり求心力を示している。双方の力が質点を安定にする。この回転事象が実在の質点を安定にしている。

2. 質点の安定化

x軸上の点（b）に質点があり、その場の存在確度が点（a）より低い場合は点（a）からの中心力が点（b）に影響するから質点はx軸上を点（a）への運動を起こす。点（b）は点（a）の偏位（X_0）となる。その質点がx軸に沿って移動している式を求めるには式（13）を因数分解して次の式を得る。

$$(19) \qquad [d/dt+(\tau/\sigma)][d/dt-(\tau/\sigma)]X = 0$$

$$(20) \qquad dX/dt-(\tau/\sigma)X = 0$$

式（20）は無限に拡散してゆく式で、ここでは現実的ではない。これは安定化の過程であり $[dX/dt+(\tau/\sigma)X = 0]$ が適用されるべきである。よって式（19）の解は次式となる。

$$(21) \qquad dX/dt+(\tau/\sigma)X = 0$$

この式の解は次の減衰関数で、X_0は初期値で点（b）の位置を示す。

$$(22) \qquad X = X_0{*}\exp[-(\tau/\sigma)t]$$

偏位は減衰係数（τ/σ）により 0 に収束する。このことは波動関数（18）の波高は 0 に収束する事を示す。これだけでは存続性が作用するから現実の説明にならない。

　質点は式（21）に示す変動性に基づき運動する。そして式（17）に示す存続性に基づき安定する。実在性に基づきこれら二つの式の和が運動を示す。

　式（21）は点（b）から点（a）まで変動性により移動する質点を表す。式（21）は次のように変更できる。

(23)　　　$2\omega dX/dt + 2\omega^2 X = 0$

存続性も運動を制御する。式（18）を下式の形にする。

(24)　　　$d^2X/dt^2 + \omega^2 X = 0$

式（10）により運動の実在性は式（23）と式（24）の和であり、下式となる。

(25)　　　$d^2X/dt^2 + 2\omega dX/dt + 3\omega^2 X = 0$

上式での X の解は次式となる。

(26)　　　$X = \exp(-\omega t) * [X_1 * \cos(2^{1/2}\omega t) + iX_2 * \sin(2^{1/2}\omega t)]$

X は実数でなければならないから $X_2 = 0$。

　$t = 0$の時、X は初期値の X_0、よって $X_1 = X_0$。

　減衰波動関数の下式（27）が得られる。

(27)　　　$X = X_0 \exp(-\omega t)\cos(2^{1/2}\omega t)$

波高 X は次第に減少してほとんど0になり、その質点は安定化する。

　式（14）と（18）は質点の De Broglie による物質波を示している。これらの式は波動関数で物質波を表し、その波長（λ）は $\lambda = h/P$。光は波であるが、粒子の特性もある、そして粒子も波の特性を有している。電子の波長は可視光線の波長の千分の一以下である。粒子や質点には隠れた波動がある。運動粒子の物質波は進行波となる。

3．空間の安定化

　実在空間は概念的空間、あるいは参照枠内の存在である。実在空間内の点には質量はない。しかし、そこには位置がある。その位置には概念

空間、あるいは参照枠の対応する点に対して種々の偏位や揺動を伴っている。実在空間内の時間が存在を変化させる。その空間内の点は不安定であるから存在確率や持続確率を伴い、それらが存続性や変動性の起因となる。ゆえにこの空間には質量のない実在性がある。実在性の点の変動性は式（13）に従い遠心加速度を持ち、また、存続性は式（18）に従い求心加速度を持つ。この２式が空間の存在を安定にする。この空間内の点は安定で回転している。実在空間内の点は安定であるから実在空間も安定である。

F. 旋回と生起率

1. 空間と線上の生起率

　質点が一空間内の如何なる点上でも存在（生起）する確率は均一である。よって、空間内の一直線上での質点の生起率も均一となる。x軸上の生起率P_Cはどの位置でも一定であるから、x軸上の一質点の偏差の確度分布は正規分布を成し、式（1）となる。単位質量（エネルギー）当たりの線生起率をP_Cとし、ある質点のエネルギーをEとする。この質点の存在の確度分布は正規分布を成し、その標準偏差値（散在性）と生起率との関係は式（4）から$\sigma^2 \fallingdotseq P_C/E$、よって線生起率と質量の比の平方根が標準偏差値であり、質点の散在性である。質点が空間内の如何なる点にも生起率は均一であるから、単位エネルギーが空間内に生ずる平均分散をε^2と置くと、エネルギー（E）の単位容積内生起率は$E*\varepsilon^2$である。また、線生起率（P_C）との関係はx軸近傍の断面積をΔsとすると、

$$(28) \qquad P_C*dx = \Delta s*E*\varepsilon^2*dx$$

$$E*\sigma^2 = \Delta s*E*\varepsilon^2$$

$$\sigma^2 = \Delta s*\varepsilon^2$$

Δsは面積の次元を持ちσ^2は距離の二乗の次元を持つからεは次元を持たない。線生起率と空間生起率の比は、

$$(29) \qquad \Delta s = \sigma^2/\varepsilon^2$$

2. 他の質点の生起率

　質点（A）は空間内どこでも生起率は一定であるから、線生起率で考えるなら x 軸上どこでも一定の線生起率（P_C）である。だがもう一つの質点が原点（0, 0）に存在していて、これに対して x 軸上の点（x, 0）に質点（A）が現れる割合は一定ではない。その生起率は距離とともに減少してゆく。まず、x 軸上の点（x, 0）に存在する質点（A）から見て原点が見える割合（R_O）は質点（A）から原点への視角の成す円錐の底面と原点を中心とする半径 x の球面との比である。視角は角速度である τ と σ の比、すなわち振動角速度（ω）であり、これの成す円錐底面と球面積の比は、

$$(30) \qquad R_O = \pi(a\omega x/2)^2/(4\pi x^2) \quad a は比例定数$$

$$= (a\omega)^2/16$$

となり、x に関係なく一定となる。また、原点から半径 x の位置に質点（A）が生ずる割合は点（x, 0）近傍の断面積 Δs と半径 x の球体表面積 $4\pi x^2$ との比となる。よって、原点から見た点（x, 0）近傍内への出現割合を R_{OX} とすると、

$$(31) \qquad R_{OX} = b\Delta s/(4\pi x^2) \qquad b は比例定数$$

$$= (b\sigma^2)/(\varepsilon^2 * 4\pi x^2)$$

ゆえに、点（x, 0）に存在する質点（A）の原点に対する生起率（R_X）は、

$$R_X = R_O * R_{OX}$$

$$= (a^2\omega^2/16) * (b\sigma^2)/(\varepsilon^2 * 4\pi x^2)$$

$$= (a^2\omega^2*b\sigma^2)/(G*x^2) \quad G = (8\varepsilon)^2\pi/(a^2b) \text{ とする}$$

(32)　　　　$R_X = \tau^2/(G*x^2)$

点 $(x, 0)$ に存在する質点（A）の原点に対する生起率は τ^2 に比例し、x^2 に反比例する。

3．他の質点への影響

　質点（A）が現実の空間内に単独で存在する場合、その生起率はどこでも一定であり、どこにでも安定して存在し得る。x軸上の点 $(x, 0)$ に存在する質点 A はその場所に安定に存在し得るが、空間内にもう一つの点が存在する場合はお互いの影響があり質点 A の存在は安定でない。点 $(x, 0)$ に存在する質点 A が原点に存在する質点 O に及ぼす影響について考えると、お互いの距離が離れているほど影響は少ない。質点 A が最も安定に存在し得る距離は無限遠で、一点の存在に同じである。また、原点から見て質点 A が x 軸上の点 $(x, 0)$ に存在する生起率は式（32）の如く x の二乗に反比例する。よって二乗が x に反比例する確率変数 (u) を考え、下式（33）の如く定義する。

(33)　　　　$u^2 = 1/x$

(u) が 0 の時は (x) は無限大でなければならない。(x) の値が x から無限大に増大してゆく場合は、u は u から減少し 0 になる。この場合双方の積分値は同じになる。x の生起率は式（32）が示す如く減少する。x における減少量は $(R_X/x^2)*dx$ で、これを積分する。結果は $-\tau^2/(2G*x)$ で、式（33）により $-(\tau^2u^2)/(2G)$ となり u に関する生起率は一定と思われ、これを q とする。u に関する増加率は $q*u*du$ であり、x に関する減少率は $-R_X*dx$ である。

$$(34) \qquad -R_x * dx = q * u * du$$

x については無限大から x まで、u については 0 から u まで両辺を積分する。

$$(35) \qquad \tau^2/(2G * x) = q * u^2/2$$

$u^2 = 1/x$ であるから $q = \tau^2/G$ となり、u に関する生起率は一定となる。

　式 (33) の如く関数 (u) を定義すると、質点 (A) は (u) を確率変数として正規分布を成す。変数 (x) が無限大の時に最も安定、即ち存在の確度が最も高くなる。関数 (u) については $u = 0$ の時に存在の確度が最大となる。平均値に関しても、如何なる u の値に関しても生起率が一定であれば、平均値は 0 である。よって、質点 (A) の偏差の確度分布は次式を成す。

$$(36) \qquad \Phi = \exp[-u^2/(2\eta^2)]/[\eta(2\pi)^{1/2}]$$

η は標準偏差値である。この式に式 (35) を代入すると次式を得る。

$$(37) \qquad \Phi = \exp[-1/(2x\eta^2)]/[\eta(2\pi)^{1/2}]$$

これが x 軸上の存在の確度分布である。持続の確度分布は揺動 0 を中心とした正規分布であり、式 (6) に従う。

$$(6) \qquad \Psi = \exp[-V^2/(2\tau^2)]/[\tau(2\pi)^{1/2}]$$

この2式に安定性理論の存続性を応用すると、

$$(38) \qquad d^2x/(dt)^2 = -(\tau^2/\eta^2)/x^2$$

となり、原点方向に距離の二乗に反比例した加速度を与える。

　$x = 1$ の場合、式 (32) から $R_1 = \tau^2/G$

式（4）により、$\eta^2 = R_1/E$

$$\eta^2 = \tau^2/(G*E)$$

この式を式（38）に代入すると次式を得る。

$$(39) \qquad -d^2x/dt^2 = (G*E)/(x^2)$$

よって点 $(x, 0)$ に存在する質点（A）は、原点方向に加速度を得て万有引力と同じくエネルギー、即ち質量に比例して距離の二乗に反比例する求心力を得る。

4．旋回

式（38）は存続性による負の加速度の存在を示す。式（38）の揺動性と散在性の比は角速度であるから点 (a) の質点は旋回することを示す。

また、式（37）と式（6）に変動性を適用すると、次式を得る。

$$(40) \qquad d^2x/dt^2 = (\tau^2/\eta^2)/x^2$$

正の加速度は遠心力を生じる。存続性は式（38）が示す如く求心力を生じるので質点は回転中心との距離を変えずに旋回する。

G. 引　力

1. 復元力

　質点の存在期待位置が点（0, 0）であるとして、その標準偏差値が比較的小さく、即ち高精度で期待値上での存在確度が高い確実な存在や、精度が悪く、その存在確度の低い存在がある。前者は少しのずれで存在確度が大きく低下する。元の位置の存在確度が高いので、戻る確率が高い。偏位が起こり難い。即ち期待値に向かう力がある。強い復元力が働く。後者は少しのずれでは存在確度の減少が比較的少ない。即ち精度が低く、その期待値に向かう力、復元力は比較的弱い。確実な存在状況とは復元力が強く物が常に所定の位置に確実に存在している状態である。不確実な存在状況とは復元力が弱く物がばらついてしまう状況で、物が本来の位置にある確率が比較的低い状態である。この場合は精度の低く散在性の高い状態である。期待値に向かう復元力を求心力と定義する。

2. 偏位と求心力

　静止状態では周囲に求心力が均等に存在するから静止を保っている。質点に小さなずれが生じたとする場合、存在頻度即ち存在確度は下がる。中心の方が高い存在確度なので復元する力が働く。存在確度の高い存在では求心力が大きく、存在位置がずれる可能性が少ない。存在確度の低い存在では求心力が小さいから位置がずれる可能性がある。存在確度の高い存在には強い求心力が伴っていて、小さなずれの力では動かない。存在期待値上の求心力が最も強く辺縁に行くほど存在確度の低下に伴い弱くなる。求心力は存在確度の、即ち偏位 X の関数である。存在期待値上では（$1/\sigma$）に比例するとする。存在期待値上の求心力は中心

力と定義する。式（2）から中心力（F）は散在確度$1/[\sigma(2\pi)^{1/2}]$に比例する値とするゆえ、

$$(41) \qquad F = E/[\sigma(2\pi)^{1/2}]$$

　Eは比例定数でエネルギーの次元を持つゆえ、その質点のエネルギーと考えられる。その求心力は偏位Xの関数で引力の原因となる。偏位XにエネルギーKがあれば、エネルギーEの偏位Xにおける存在確度分布によりエネルギーKに求心力を掛ける。同時にエネルギーKはX＝0の点にKの存在確度によるKに向かう求心力を起こす。この両者の積が引力となる。

3．万有引力

　単位エネルギーの質点が原点に存在すると、その近傍に散在性（σ_1）の存在確度分布、$\exp[-X^2/(2\sigma_1^2)]/[\sigma_1(2\pi)^{1/2}]$をもたらし、その偏位X上にエネルギーKの質点が存在した場合、質点KはX＝0への求心力（$-F$）を得る。単位エネルギーがエネルギーKにXの位置でもたらす求心力は、

$$(42) \qquad -F = K\exp[-X^2/(2\sigma_1^2)]/[(2\pi)^{1/2}\sigma_1]$$

エネルギーKが原点にもたらすエネルギー頻度がこの求心力を増加させる。点Xに存在するエネルギーKの存在確率分布が原点に存在するエネルギーEの存在頻度を規定する。それは$E\exp[-(-X)^2/(2\sigma^2)]/[(2\pi)^{1/2}\sigma]$であり、点Xに存在するエネルギーKの求心力を増加させる。

　このエネルギー頻度と式（42）の積が求心力（$-F$）となる。

$$-F = E\exp[-(-X)^2/(2\sigma^2)]/[\sigma(2\pi)^{1/2}]$$

$$*K\exp[-X^2/(2\sigma_1^2)]/[\sigma_1(2\pi)^{1/2}]$$

$$= KE\exp[-(-X)^2/(2\sigma^2)-X^2/(2\sigma_1^2)]/(2\pi\sigma_1\sigma)$$

$$= KE\exp[-X^2(\sigma^2+\sigma_1^2)/(2\sigma^2\sigma_1^2)]/(2\pi\sigma_1\sigma)$$

X の値が十分に大きい場合は、テイラー展開を応用して近似式にする。

$$(43) \qquad -F = KE[2\sigma^2\sigma_1^2/X^2(\sigma^2+\sigma_1^2)]/(2\pi\sigma_1\sigma)^2$$

$$= KE[\sigma\sigma_1/(\sigma^2+\sigma_1^2)]/(\pi X^2)$$

ニュートンの万有引力の式を得る。

$$-F = KE\{\sigma\sigma_1/[\pi(\sigma^2+\sigma_1^2)X^2]\}$$

$$(44) \qquad -F = G*KE/X^2$$

G は引力定数であり、$G = \{\sigma\sigma_1/[\pi(\sigma^2+\sigma_1^2)]\}$ である。

　E と K との引力は E の K にもたらす存在頻度と K の E にもたらす存在頻度の積である。

H. 相 対 性

1. モメンタムとエネルギー

ある質点が一つのユークリット空間内で、ある点上に安定して静止している場合は速度0である。質点がある点上に確実に存在していることは質点が一点上に持続して存在していることであり、静止（速度0）していることである。質点が静止していることは持続確度分布で揺らぎなし（揺動速度0）の確度が最大で、その近傍の揺動はより低い確度で分布していることである。揺らぎ難い存在とは静止のはずみが大きく堅固な状態の存在である。τはΨの標準偏差で揺動性を表すが、これが小さいほど揺動速度 V＝0の確度が高くなり強度が増す。この標準偏差値の逆数（1/τ）は堅固さに比例するゆえ強度と呼ぶ。

エネルギー（E_1）と質点の持続確度の積、即ち、点 a に静止している質点のエネルギー（E_1）と式（6）の V＝0のときの確度$1/[\tau(2\pi)^{1/2}]$ との積（P）はモメンタムの次元になる。

$$(45) \qquad P = E_1/[\tau(2\pi)^{1/2}]$$

よって、モメンタム（P）は強度に比例する、比例定数 E_1 はエネルギーの次元である。強度が高いほど静止モメンタムも大きく安定である。

モメンタムと角速度の積は力であるから、中心力（F）と静止モメンタム（P）の比（F/P）も角速度（ω）である。FとPの比は式（41）と（45）から次式となる。

$$F/P = [E(2\pi\sigma^2)^{1/2}]/[E_1/(2\pi\tau^2)^{1/2}]$$

$$= (\tau/\sigma)/(E_1/E)$$

（F/P）も（τ/σ）も（ω）であるゆえ、$E_1/E = 1$となる。Eは散在性にも揺動性にも共通であり、静止モメンタムや中心力の大きさに係わる状態を規定する。また、その質点が代表する物体のエネルギーである。

2. 速度と相対性

　空間α内に一質点が静止している空間βが含まれており、その静止エネルギーをEとし、その静止モメンタムをP_0 [$= E/(2\pi\tau^2)^{1/2}$、式（45）より]とする。空間βが空間αの中を速度vで運動している時、空間βの中ではエネルギーもモメンタムも変化しない。それらは静止状態である。空間αの中ではモメンタムは変化するが静止エネルギーは変化しない。空間α内を運動している質点のモメンタムは上昇する。揺動性の静止時からの変化による。（τ′）を運動中の揺動性とする。それは質点が空間α内を運動している方向における持続確率分布の標準偏差値である。式（45）から静止モメンタム（P_0）は$E/(2\pi\tau^2)^{1/2}$で、速度（v）で運動しているモメンタム（P）は$E/(2\pi\tau'^2)^{1/2}$であり、モメンタム（P）は速度と共に上昇するので揺動性（τ）から（τ′）まで減少するが、静止エネルギー（E）は変わらない。速度（v）で運動している質点の揺動（V）の確度分布は運動速度（v）の影響を受け、揺動速度の平均は0ではなく速度の方向にシフトする。シフトの平均を速度（v）にある関数（f）を乗じたものと置く。この場合の揺動性を示す標準偏差値は（τ′）で、その二乗である平均分散（τ′）^{2}は揺動確度分布の分散（V$-$$fv$)^{2}の平均である。$fv$がτに近い範囲では、この平均分散は式（6）における静止時の平均分散（τ^2）から運動中の（fv）の二乗を減じたものとなる。

$$(46) \qquad (\tau')^2 \fallingdotseq \tau^2 - (fv)^2$$

静止している時は$v = 0$であり、（τ′）^{2}とτ^{2}は一致する。vが大きくなる

にしたがって $(\tau')^2$ の値は小さくなるが $(\tau')^2$ の値は 0 または正数でなければならないから、$(fv)^2$ の最大値は τ^2 である。ゆえに v の最大値を C とすると、

$$(47) \qquad \tau^2 = (f\mathrm{C})^2$$

$$(48) \qquad f^2 = \tau^2/\mathrm{C}^2$$

f は定数となるので次式を得る。

$$(49) \qquad (\tau')^2 = \tau^2[1 - (v/\mathrm{C})^2]$$

質点の運動速度には限界があり、この限界の速度は光速またはそれに近い速度と考えられる。

　相対性理論ではこれを光速と考える。運動している質点のモメンタムを求めるには、式 (45) より質点の静止モメンタム (P_0) は $\mathrm{E}/(2\pi\tau^2)^{1/2}$ であり、速度 v の時のモメンタム (P) は $\mathrm{E}/(2\pi\tau'^2)^{1/2}$ であるから、次式を得る。

$$(50) \qquad \mathrm{P} = \mathrm{P}_0/[1 - (v/\mathrm{C})^2]^{1/2}$$

速度が上昇するにしたがってモメンタムは無限に大きくなり得るが、これは速度が限界に近づくからである。この限界速度が光速だと考えられる。

3. 質量とエネルギー

　静止している質点には持続確度分布があり、それに対応しているエネルギーがある。質点が運動するとその運動速度に伴って揺動性は減少し、エネルギー (E) が増加する。速度には限界があり、最高速度を (C) とする。限界速度でエネルギー増加はなくなる。

式（45）の両辺に最高速度 C を乗じる。CP の次元はエネルギーである。

$$(51) \qquad CP = CE/[\tau(2\pi)^{1/2}]$$

CP は全エネルギー（E）である。

$$(52) \qquad E = CP$$

E_0 を静止エネルギー、E を運動中の質点の総エネルギーとする。
式（52）から $E_0 = P_0*C$。式（50）を式（53）の如く変更する。

$$(53) \qquad E = E_0/[1-(v/C)^2]^{1/2}$$

式（53）の近似式は次式である。

$$E = E_0*[1+(v/C)^2/2]$$

$$= E_0+(P_0/C)v^2/2$$

よって、運動エネルギーは $(P_0/C)v^2/2$ となる。これがニュートンの運動エネルギー（$m_0v^2/2$）と同じである。

$$E = E_0+m_0v^2/2$$

よって、 $\qquad m_0 = P_0/C$
式（50）の両辺を C で割る。

$$P/C = (P_0/C)/[1-(v/C)^2]^{1/2}$$

静止質量（m_0）は $m_0 = P_0/C$ であるから運動中の質点の質量（m）は $P/C = m$。
式（52）から $E/C = P$、よって $E/C^2 = m$。

式（53）の両辺を C^2 で割る。

$$E/C^2 = (E_0/C^2)/[1-(v/C)^2]^{1/2}$$

総エネルギーを最高速度の二乗で割ると総質量となり、式（54）が得られる。

(54)　　　　$m = m_0/[1-(v/C)^2]^{1/2}$

m は総質量、m_0 は静止質量

Ⅰ. 不確定性と変動性

(1) モメンタムの変動性

事象には常に散在性と揺動性がある。その散在性や揺動性が事象の変動性や存続性をもたらしている。作用反作用の法則はモメンタムの変動性による。モメンタムにも散在性や揺動性がある。モメンタムを微分すると従属変数は力となる。よってモメンタム持続確率の統計変数は力である。モメンタムと力が変動性と存続性を構成する。モメンタムの変動性は次のように求められる。

モメンタム（P）の存在確率分布（Φ）は標準偏差を（σ_p）として、（$P = P_0$）の点の周りでは式（55）の如くとなる。

$$(55) \qquad \Phi = \exp[-(P-P_0)^2/(2\sigma_p{}^2)]/[\sigma_p(2\pi)^{1/2}]$$

σ_pはΦの標準偏差値

P_X を（P_0）からの偏位とすると、$P_X = P-P_0$で、その分布は次式の如くである。

$$(56) \qquad \Phi = \exp[-P_X{}^2/(2\sigma_p{}^2)]/[\sigma_p(2\pi)^{1/2}]$$

揺らぎは偏倚（P_X）の時間的な変化であるから、力の次元である。

$$(57) \qquad F = dP_X/dt$$

この揺らぎの式のF＝0の場合は慣性の法則と呼ばれている。モメンタムの持続確率分布（Ψ）の変数は力（F）となる。

$$(58) \qquad \Psi = \exp[-F^2/(2\tau_p{}^2)]/[\tau_p(2\pi)^{1/2}]$$

変動性においては散在性と揺動性が比例している。変動性（Z）は式

（7）のように存在確度（Φ）と持続確度（Ψ）の商であるから式（56）と式（58）とを式（11）に代入し、処理すると次式（59）を得る。

$$（59）\qquad (dP_X/dt)P/\sigma_p{}^2-(dF/dt)F/\tau_p{}^2 = (dZ/dt)/Z$$

P_X は p 軸上の存在位置（P_0）からの偏位であり、F は $F = dP_X/dt$。
　変動性（Z）は定数であるから次式を得る。

$$d^2P_X/dt^2/\tau_p{}^2-P_X/\sigma_p{}^2 = 0$$

$$\left[dP_X/dt+(\tau_p/\sigma_p)\right]\left[dP_X/dt-(\tau_p/\sigma_p)\right]P_X = 0$$

$dP_X/dt-(\tau_p/\sigma_p)P_X = 0$ は拡散するので、$\tau_p/\sigma_p = \omega_p$。上式の解は次式となる。

$$（60）\qquad dP_X/dt+\omega_pP_X = 0$$

この式は作用反作用の法則の説明となる。モメンタム P_X は作用力 ω_pP_X であり、そこに存在する反作用の力は$-dP_X/dt$ である。

⑵ 不確定性原理

　物理的な事象には散在性や揺動性を伴う。散在性や揺動性のない確率は存在しないので揺動性のない事象は存在しない。ばらつきを伴う散在性とふらつきを伴う揺動性が事象の永続性と変動性をもたらす。事象は変動する。不確定性原理は一般に $\Delta E*\Delta t \geqq h$ または $\Delta P*\Delta x \geqq h$ と表示される。この二つの表示法は同一である。エネルギーの距離微分 $\Delta E/\Delta x \fallingdotseq F$ は力であり、モメンタムの時間微分 $\Delta P/\Delta t \fallingdotseq F$ も力である。よって、$\Delta E*\Delta t = \Delta P*\Delta x$。双方の表示は同じである。ここでは、この原理を説明するのにモメンタムを用いる。モメンタムにも存続性と変動性がある。モメンタムが P_0 から P に増加している時、その増加カーブは変動性による減衰指数関数である。変動性を用いてハイゼンベルグ

による不確定性原理が説明される。

⑶ 位置と運動の不確定性

不確定性原理は一般に次式の如く表記される。

$$(61) \qquad \Delta x * \Delta P \geqq n * h \qquad n \geqq 1$$

$\Delta x * \Delta P$ を微分する。

$$d(\Delta x \Delta P)/dt = \Delta x * d\Delta P/dt + \Delta P * d\Delta x/dt$$

Δx と ΔP に変動性を応用すると、$d\Delta x/dt = -\omega_1 \Delta x$、$d\Delta P/dt = -\omega_2 \Delta P$ であるから、

$$(62) \qquad d(\Delta x \Delta P)/dt = -\omega_1 \Delta x \Delta P - \omega_2 \Delta x \Delta P$$

$$= -(\omega_1 + \omega_2) \Delta x \Delta P$$

$$= -\omega_3 \Delta x \Delta P \qquad \omega_3 = \omega_1 + \omega_2 とする。$$

Δx と ΔP は正の値であり、$\Delta x \Delta P$ 減少関数と考え得るので、$d(\Delta x \Delta P)/dt$ はエネルギーの次元であるが負の値である。エネルギーはエネルギー粒子 $h\omega$ の n 倍であるから、

$$d(\Delta x \Delta P)/dt = -nh\omega$$

$$-\omega_3 \Delta x \Delta P = -nh\omega$$

ω と ω_3 はいずれもエネルギー $d(\Delta x \Delta P)/dt$ の角速度であるから同一であり、

$$\Delta x \Delta P = nh \qquad n \geqq 1$$

ゆえに、

$$(63) \qquad \Delta x \Delta P \geqq h$$

⑷ 質点と時間の不確定性

　頑丈な物体でもいずれ変動する。それには少量のエネルギー消失（ΔE）を伴う。少量のエネルギー消失での大きな変化でも短い時間が必要で、変化時間（Δt）は0ではない。変化にはエネルギー（ΔE）と時間（Δt）を必要とする。いずれも0ではない。このことは量子レベルでの不確定性原理として知られている。ΔE*$\Delta t \geqq h$ は次のように求められる。

　モメンタムの微分は力である（$dP/dt = F$）。ここでは $\Delta P/\Delta t$ を dP/dt の近似値とする。よって、

$$\Delta t \fallingdotseq \Delta P/F$$

式(45)は、　　　　$P = E/[\tau(2\pi)^{1/2}]$

上式から、　　　　$\Delta P = \Delta E/[\tau(2\pi)^{1/2}]$

式(41)は、　　　　$F = E/[\sigma(2\pi)^{1/2}]$

よって、　　　　　$\Delta t = \Delta E/E/\omega \quad \omega = \tau/\sigma$

$$\Delta E \Delta t = (\Delta E/E) * \Delta E/\omega$$

Eが大変小さくてΔEに近い場合には、$\Delta E/E \fallingdotseq 1$。

　$\Delta E \fallingdotseq nh\omega$ であるから、

$$(64) \qquad \Delta E \Delta t \fallingdotseq nh$$

よって、$\Delta E \Delta t \geqq h$

しかし E が大きい場合は、

$$\Delta E / E \fallingdotseq 0$$

よって、

$$\Delta E \Delta t \fallingdotseq 0$$

　この法則は質量が大変小さい場合に、おそらく量子レベルに適応できる。

J．仮想的存在

1．仮想的エネルギー

　式（2）における標準偏差値が散在性である。この標準偏差値が増加するにつれて質点の存在確率は減少し、標準偏差値が無限大になれば質点の実質は消失する。式（41）が示す如く求心力も消失する。これが仮想状態への移行である。質点のエネルギーを示す式（41）におけるＥの存在確度分布は平坦となる。また、散在性が負となる仮想的存在も考え得るのである。そこではエネルギーＥが仮想的存在を代表し、式（41）が示す如く遠心力が働く。エネルギー分布には山や谷ができることが考えられる。散在性は負の値であるから式（41）からわかるように、エネルギーの山（Ｅ）には遠心力が働き崩れやすく、谷（−Ｅ）には求心力が働き埋まり易い。

2．仮想状態の力

　仮想質点に式（2）を適用する。存在確度は負になるが仮想質点は式（2）に当てはまる。そのＸは仮想質点、即ちエネルギーの山または谷の中心からの偏位である。

$$\text{（2）} \qquad \Phi = \exp[-X^2/(2\sigma^2)]/[\sigma(2\pi)^{1/2}]$$

エネルギーの山が為す遠心力は標準偏差の絶対値が小さいほど強力である。エネルギーの谷はエネルギーを引き寄せる。だから谷は平らになる傾向がある。山は崩れる傾向がある。式（41）をエネルギーの山に当てはめればＦは負になり、遠心力を表す。山の中心点（Ｘ = 0）の虚弱さは $1/[\sigma(2\pi)^{1/2}]$ である。この絶対値が大きいほど虚弱さは大きい、即ち

平らになる力が強くなる。標準偏差の絶対値が減少するほど、虚弱さは増加する。だから標準偏差の逆数は中心部の負の強度、平らになる力を表す。遠心力はXの関数で、その中心はその強度に比例する。式（2）から負の中心力は$1/[\sigma(2\pi)^{1/2}]$に比例する。

$$(41) \qquad F = E/[\sigma(2\pi)^{1/2}]$$

比例定数Eはエネルギーである。Eが単位エネルギーの場合は、偏倚X におけるエネルギー分布率を表す。$-F$は中心部の平らになる力を表す。

3．実態と仮想状態

　物体はエネルギー量子から構成されている実質的な存在である。例として、中性子は実質的な存在である。中性子は陽子と電子になり得る。それらはエネルギー量子から成り立っている。中性子は陽子と電子になり得るが、双方とも完全にバランスの取れたエネルギー量子から成り立った存在に分裂するとは言えない。仮想的存在である余分なエネルギー量子が陽子に残り、エネルギー量子の不足が電子側に起こる。これも仮想的存在である。それゆえクーロン力が発生し、電子はある角速度をもって陽子の周りを回転し存続するのである。強力なエネルギー、即ち周波数の高いエネルギーは量子化または粒子化する。量子には固有のエネルギーレベルが有るのでエネルギーの過不足が起こり得る。この過不足がエネルギーの仮想状態である。

4．クーロン力

　エネルギーの山は脆く、その周囲に遠心力を発生する。エネルギーの山Eがあり、それを＋のクーロンの塊と考えると遠心力を持つ。エネルギー塊中心部のエネルギーの存在頻度は$E/[\sigma(2\pi)^{1/2}]$であり、その

遠心力は式（14）を用いて得られる。もし E を中心とする存在確度分布の偏位 X に K のエネルギーで＋のクーロンの塊があったら、中心に存在するエネルギー E の遠心力にも変化を起こす。この場合、X ＝ 0 の位置のエネルギー存在頻度はエネルギー E の存在頻度とエネルギー K の存在頻度の積である。散在性（σ）は共通であり、エネルギー E と K のクーロン力（F）は反発する力である。

$$F = E\exp[-0^2/(2\sigma^2)]/[\sigma(2\pi)^{1/2}]*K\exp[-X^2/(2\sigma^2)]/[\sigma(2\pi)^{1/2}]$$

$$= KE\exp[-X^2/(2\sigma^2)]/(2\pi\sigma^2)$$

X が十分大きければ式（43）の如くテイラー展開を用いて近似値を得る。クーロン力の式が得られる。

$$F = KE[2\sigma^2/(-X)^2]/(2\pi\sigma^2)$$

$$= KE/(\pi X^2)$$

分布の中心（X ＝ 0）にエネルギーの谷が在ったとしたなら、そのエネルギーは負のクーロンであり求心力となる。

$$-F = KE/(\pi X^2)$$

負のクーロンは正のクーロンを引き寄せる。エネルギー E とクーロン（q）の関係は次式となる。

$$(65) \qquad q = E/\pi^{1/2}$$

K. エネルギー波

1. 空間の不安定性

　宇宙の大空間が如何なるものであるかは把握困難ではあるが、その存在の概念はある。空間は一般的に多次元の概念的な存在である。しかし物の存在する三次元空間は実在する。ある一点を基準として基準3方向を規定して得る空間には二次空間が実在し得る。またこの二次空間は動かし得るゆえ不安定な存在である。よって変動性と存続性が存在する、即ち物理的な実在性がある。それゆえ現実の空間の全ての点には散在性と存在確率、揺動性と持続確率がある。位置は正確で不安定性もない概念上の空間と物理的に存在する現実の空間とは明確な区別がある。物理的存在は一般的に質量の存在により認識されるが質量の無い存在も認識されなければならない。

　実在空間内に存在する二次空間はいかに小さくても実在性がある。空間内の一点を取り巻く微小空間は実在する。この点に、その実在微小空間を代表させ実在点とする。点そのものは概念的存在であるが、実在点とはその点を中心とする微小空間であり、その大きさは特定できないが特定微小空間の集合体でもあり得る。実在空間内の全ての実在点やその集合体は回転、即ち波動していることを式（14）と（18）が示す。実在空間は潜在エネルギーである。実在空間は全ての実在物の源である。

2. エネルギー源と波動

　空間は潜在エネルギーである。種々の可能性を持つ物理的に存在する実在点には位置と運動がある。それらには不安定性もあって偶然性が作用する。偶然性はそれらの実在点を集合させ集合体を形成する。これら

の集合体はそれぞれ存在確率分布や持続確率分布を持つ。それら集合体には色々な散在性や揺動性がある。それらの集合体には色々な実在性があるから多様な強度で回転している。その回転強度は偶然に起きる集合体の大きさによる。エネルギーの基は実在点の集まり、あるいは圧縮された点の集合体である。大きな集合体は強いエネルギーを造る力がある。これらの集合体はポアソン分布で分布する。よって大部分は弱い力の集合体である。

この空間には温度がある。それはエネルギーであり、弱い力の集合体エネルギーで満たされていることである。強力な集合体は多くはない。これらの集合体は動き回って波となる。それがエネルギーである。物理的な空間は色々な強さのエネルギー波で満たされている。散在性の高いエネルギー源は散っていく。揺動性の高いエネルギー波はそのエネルギー形態を変化させ易く中間子や粒子に成り易い。

式（14）と式（18）が示すように、実在空間は回転して集合体を形成する点から形成されている。散在性の高い集合体は散りやすい。散在性の低い集合体は簡単には散らない。

散在性や揺動性に基づいてこれらの集合体は動く。動くときは波となる。式（13）が示す如く周波数が低く揺動性の低い集合体は散在性が高い。この種のエネルギーは散りやすい、即ち熱線である。周波数の高い集合体は揺動性が高く散在性が低い。それらは変化し易く散り難い。それらが集まるとその力が上昇する。それらは強力な集合体となり色々な高震度のエネルギー波となり、数珠状に重合する。それらの重合体がエネルギー量子となり光線となる。大変強力な集合体は大変高震度の波があり揺動性は大変高く散在性は大変低い、ということは殆ど広がらない。それらは回転となり大変小さな粒子となって集合し、質量の有るまたは無い量子となる。それらがクオークやメゾンである。

集合体は集成したり崩壊したりを繰り返すエネルギーの源である。エネルギーが低下することは起こり得る。強力な集合体発生確率はポアソ

ン分布においては低い。強力な集合体が弱い集合体に崩壊し得る。弱い集合体は実在空間の実在点となり得る。潜在的にエネルギーの総量は変化しない。

3. エネルギー

　エネルギーは最も基本的な存在である。実在空間のすべての実在点には対応する概念空間の点に対して不安定性があり集合体を形成する。実在空間の集合体には散在性や揺動性があり、式（14）と（18）からわかるようにそれらは回転している。回転して散る各集合体は波となる。波はエネルギーである。実在空間は潜在的なエネルギーである。そこに存在するエネルギーの基がエネルギーの力を規定している。強力なエネルギーには高度の潜在的可能性が必要である。強力なエネルギーの発生率は低く、低エネルギーの発生率より低いと考えられる。なぜならエネルギー源の強さの分布はポアソン分布によると考えられるからである。エネルギーには量と強さがある。量は振幅による。大海の大きな波のエネルギー量は大きい。砂浜の小さな波のエネルギー量は小さい。エネルギー量、即ち振幅の違いである。強さは振動数による。強さの低いエネルギーは熱である。散在性の高い集合体からの波は広がり易い、それは熱である。揺動性の高い集合体からの波は振動数が高く（式27）強力であり、エネルギーの種類を変え易い。光量子はかなり強力である。強力なエネルギーは揺動性が高いので存続性も高く質量を形成する。高振動は粒子化しクオークや中間子等、質量の有る粒子や無い粒子に変化する。光粒子や質点はエネルギー担体、即ち運搬車で、この担体は位置や運動があり、それなりの実在性がある。時間が実在空間内で最も重要な要素である。実在空間はエネルギーの波動で満たされている。波はエネルギーである。

4．熱

　熱は実在している。よって散在性と揺動性から成り立っている。だから変動性も存続性もある。熱の角速度はそれほど高くはない。熱はエネルギーであり、その波の角速度は大変高いわけではない。熱エネルギーは高い散在性があるので変動性は少ない。簡単には変わらない。熱は空間内で散らばり易く存在確度は低い方であるが、決して平らではない。だから散在性（σ）は大きいが限度はある。熱はどの方向へでも浸潤する。よって揺動性（τ）は小さい。エネルギーは波であるから波動関数の式（18）が示す振動数はむしろ遅い角速度を示すであろう。角速度はエネルギーの強さを表すのであるが、これが高くなれば量子化する。そして大変高い角速度は粒子化するし、質量を持つ粒子になれば物質となる。

5．光

　周波数の高い波は重合して数珠状になる。これが光量子である。光は実在性のある量子化した存在である。光量子はエネルギー波から成るエネルギー担体、即ち運搬車である。光量子も位置と運動があり存在確率や持続確率がある。よって散在性や揺動性があり、変動性や存続性もある。光量子の揺動性は大変高く変動性が高い（式14）ので直ぐに消失する。持続しない、持続確率は大変低い。光量子は直ちにエネルギー波に変化する。光量子の存在確度は高く、その散在性は小さい。よって光の角速度（$\omega = \tau/\sigma$）は大きい。光量子は空間内で局在するが瞬間的に消失する。式（18）が示す存続性では光量子は高振動であることを示している。光量子には変動性や存続性がある。その存続性は0か、0に近い値である。光はすぐに消滅する。光の存在確度は高く、その散在性（σ）は小さい。光量子は空間内で局在するが一瞬である。式（18）

が示す存続性では光量子には高振動数がある。変動性の式（22）が示す
ように偏位は殆どなく一定で、変動性は殆ど0であることを示してい
る。だから光量子には変動性は殆どなく空間を一直線に進む。光は真っ
直ぐに進み周りの物に影響されない。光はエネルギー波が積み重なった
物質波が一直線内に閉じ込められたものである。光量子は極端な物質で
ある。光量子には実在性がある、その構成要素は極端な値のものである
が、なお波動の特性がある。光は量子化したエネルギーであり光量子は
物質波で、それに伴う波長は4000から8000オングストロングである。
光量子以上の振動数の量子は強く回転して粒子化する。大変強い回転で
は質量と成ると考えられる。

L．ロレンツ関数

1．ロレンツ関数と変動性

ロレンツ関数は変動性から導くことができる。次に示す式（13）は変動性から導かれた式である。

$$(13) \qquad d^2X/dt^2/\tau^2 - X/\sigma^2 = 0$$

両辺を積分する。C_0 は時間に関係しない定数である。下式上段の式の τ' は運動中の時空座標における揺動性の値で、下段の τ は基本の時空座標における揺動性の値である。

$$\sigma^2 dX/dt - \tau'^2 X^2/2 = C_0{}'$$

$$\sigma^2 dX/dt - \tau^2 X^2/2 = C_0$$

下式から上式を引く。

$$\tau'^2 - \tau^2 = 2(C_0 - C_0{}')/X^2 \quad \text{よって、}$$

$$\tau'^2 = \tau^2 - 2(C_0{}' - C_0)/X^2$$

τ'^2 と τ^2 の双方の次元は速度の二乗なので、$2(C_0{}' - C_0)/X^2$ を $(fv)^2$ と置く。f は定数である。

$$\tau'^2 = \tau^2 - f^2 v^2$$

v が増加するに従い τ'^2 は減少するが、τ'^2 は 0 または正であるから τ'^2 が 0 の時、v は最大速度 C となる。

$$f^2 = \tau^2/C^2$$

$$\tau/\tau' = 1/(1-v^2/C^2)^{1/2}$$

この式の右辺がローレンツ関数である。この関数は、質量、エネルギー、モメンタム等の補正に必要な関数であり、これを γ 補正関数と呼ぶ。γ 補正関数は1以上の値となり、座標変換等にも必要になる。

２．運動の相対性

　質点の運動を二つの座標系の相対性から考え直してみる。ある基準点 R に準拠する空間を座標系（α）と呼ぶ。この座標系（α）の x 軸方向に速度 v で移動している基準点 R′ に準拠する空間（座標系〈β〉と呼ぶ）内の一点上に存在する質点（A）の座標系（α）から見た揺動の確度分布（Ψ）を考える。質点（A）の座標系（α）上での揺動性（τ'）と座標系（β）上での揺動性（τ）は必ずしも一致しない。質点（A）は β 上では静止しているが α 上では運動しているからである。揺動の平均は β 上では0であるが、α 上では0ではなく、運動速度（v）の影響を受ける。揺動の平均値として速度にある関数（f）を乗じた（fv）とする。この分布の分散（$V-fv$）^{2} の平均値であり、それは V^2 の平均値は τ^2 であるので τ が fv の近似値であれば下式の如くである。

$$(\tau')^2 \fallingdotseq \tau^2 - (fv)^2$$

座標系（β）が座標系（α）に対して静止している時は、$v=0$ であり、（τ'）と τ は一致する。v が大きくなるに従って（τ'）^{2} は小さくなるが正または0でなければならないから、式（48）と同様に $f^2 = \tau^2/C^2$ となり次式を得る。

$$(66) \qquad \tau' = \tau/[1-(v/C)^2]^{-1/2}$$

よって、τ′＝τ/γで、τ/τ′の比はγ補正関数に等しい。座標系（α）から見た質点（A）の揺動性（τ′）は座標系（β）上の揺動性（τ）より小さくなる。揺動性は速度であり座標系（α）から見た速度は遅くなる。

3．距離の相対性

　散在性σの相対性についても揺動性と同様な議論が当てはまる。座標系（α）に対してvの速度でx方向に移動している座標系（β）上に静止している質点（A）として、この場合での座標系（α）上の質点（A）の期待値との偏位 X の散在の確度分布（Φ）を考える。質点（A）の座標系（α）上での散在性と座標系（β）上での散在性は必ずしも一致しない。座標系（α）から見た散在の確度分布は座標系（β）が運動しているという要素が入るからである。速度（v）で運動している座標系（β）上に静止している質点を座標系（α）から見たばらつき、即ち存在の確度分布の偏差の平均が０ではなく、速度（v）の影響を受け速度方向に変化するからである。偏差の平均は速度（v）にある関数（g）を乗じたものとする。座標系（α）上での標準偏差値が散在性（σ′）で、その二乗が平均分散（σ′）^{2}で、散在の確度分布の分散（X－gv）^{2}の平均である。この平均分散は式（2）における静止時の平均分散即ち座標系（β）上での質点（A）平均分散（σ^2）から偏差の平均（gv）の二乗を減じたものとなるから、

$$(σ′)^2 ≒ σ^2 － (gv)^2$$

座標系（β）が座標系（α）に対して静止している時は、（σ′）と（σ）は一致する。（v）が大きくなるにしたがって（σ′）は小さくなるが、正または０でなければならない。よって（v）の最大値を C とすると、

$$g^2 = σ^2/C^2$$

g は定数となるので次式（67）を得る。

$$(67) \qquad \sigma' = \sigma / \left[1 - (v/C)^2 \right]^{-1/2}$$

よって、$\sigma' = \sigma/\gamma$ である。質点（A）の散在性（σ）の次元は長さである。座標系（β）が静止している時は、座標系（β）上の長さ L のものは座標系（α）上でも L であるが、座標系（β）が運動していると $\sigma' = \sigma/\gamma$ となり、L も座標系（α）上では L/γ であり、「運動している座標系の長さは短くなる」。

4．時間の相対性

　一つの三次元空間の中にも幾つかの三次元空間が存在し得る。各々の空間は一つの基準点と三つの基準の方向から成り立っている。ある基準点が他の基準点に対して運動している場合もある。各々の三次元空間内の位置関係は、その空間の位置ベクトルと基準点の位置ベクトルの和で決まる。この場合、ガリレオ相対論では時間には相対性はなく各空間に共通と考えた。自然の時間即ち現時刻は常に現在であり、これは如何なる三次元空間においても共通である。時間の経過は共通ではない。過去からの経過時間を負の時間、将来への経過時間を正の時間と考えるなら、時間即ち、ある時刻から次のある時刻までの経過時間は測定の対象となる連続物理変量となる。経過に関する連続時間では過去も現在も将来も同時に考えられなければならない。いま、ここではある時刻からある時刻までの間、即ち概念的、数理的な時間を時刻差または数理時間と称する。それに対して物理的に測定される時間を自然時間と称する。自然時間（以後時間と称する）は概念的な数理時間即ち時刻差（以後時刻と称する）に対して不確実性を生じる。時間と時刻の差（T）は 0 ではなく、平均が 0 で、0 を中心にある標準偏差値を持って正規分布を成している。時間軸上の各点は対応する時刻に対して一定の標準偏差値を

持った統計的存在である。ある基準体 R に準拠する時間、即ち座標系（α）の時間軸上の一点と、その点に対応する時刻との差を T とし、その不確実性の標準偏差値を v とすると、

(68)　　　　　$\Upsilon = \exp[-T^2/(2v^2)]/[v(2\pi)^{1/2}]$

$v は \Upsilon の標準偏差$

Υ は時刻と時間の差の確度分布であり、0 である確率が最も高い。v の次元は時間である。他の基準体 R′ に準拠する座標系（β）にも式（68）は当てはまる。この基準体 R′ が基準体 R に対して静止している場合には式（68）は座標系（α）から見た座標系（β）の時間にも当てはまる。しかし、基準体 R′ が基準体 R に対して運動している場合は、座標系（α）から見た座標系（β）の時間の不確実性は上昇する。基準体 R′ が基準体 R に対して、ある一方向、例えば座標系（α）の x 軸方向に速度 v で運動している場合の時間と時刻の差（T′）は T に一致しない。T′ の平均値は速度の方向でずれる可能性がある。その平均値を速度（v）とある関数（h）との積とする。T′ の分散は T の分散とは平均値以外変化はない。よって（T′$-hv$）^{2} の平均は T の平均分散（v^2）に一致する。T′ の平均分散は（v'）であるので下式を得る。

$$v^2 \fallingdotseq (v')^2 - (hv)^2$$

$v = 0$ の時は双方の時間の不確実性は一致する。v^2 は正または 0 であり $v = 0$ の時は v が最大で値を C とすると h の値は定数となり、また $h^2 = (v')^2/C^2$ となるゆえ、

$$v^2 = (v')^2 * (1 - v^2/C^2)$$

(69)　　　　　$v' = v * [1 - (v/C)^2]^{-1/2}$

基準点 R′ が基準体 R に対して運動している場合は、v′＝v*γ であり、基準体 R に準拠する時間に対して基準体 R′ に準拠する「運動している時間は延長する」。

よって、各座標系に準拠する時間は必ずしも一致しない。個々固有の時間軸が存在する。

５．座標変換

基準体 R に準拠する座標系（α）の x 方向に速度 v で運動している他の基準体 R′ が準拠する平行な座標系（β）との座標変換を考える。

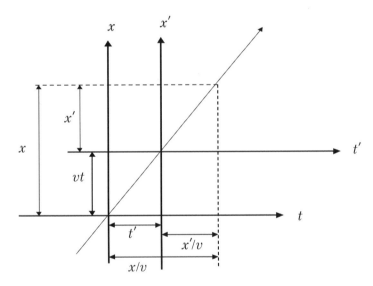

便宜上 x 軸方向のみの変換を考える。基準体 R 及び R′ はある時刻に重なっていたとする。そこから t 秒後には基準体 R′ は x 軸上を vt だけ進む。この時 x が $x′$ と重なっているとすると、上図の $x′$ は $x-vt$ と等しくなる。ただし $x′$ は座標系（α）からは γ だけ短く見えるから、

$$x'/\gamma = x - vt$$

ゆえに、

(70)　　　$x' = (x-vt)\gamma$

　運動している基準体に準拠する座標系の距離の変換には速度、時間だけでなく γ 値が関与する。また、時間軸の方を見ると、t' と x'/v の和は x/v に等しい。

　ただし、運動している座標系の時間は γ だけ長くみえるので、

$$(t'+x'/v)\gamma = x/v$$

$$vt' = x/\gamma - x'$$

$$vt' = x/\gamma - (x-vt)\gamma$$

$$vt' = [vt - x(1-1/\gamma^2)]\gamma$$

$$vt' = (vt - v^2x/C^2)\gamma$$

(71)　　　$t' = (t - vx/C^2)\gamma$

運動している基準体に準拠する座標系の時間の変換には距離、速度だけでなく γ 値が関与する。

M. 安定化原理の生物医学的応用

1. 細胞培養における安定化原理

(1) 最適細胞数

　細胞は周りの環境条件が許す限り増殖する。細胞には分裂して増殖する衝動がある。しかしとりまく生物的条件がその増殖を制限する。その生物的条件により適切な細胞数、即ち最適細胞数が決まる。細胞の適切な存在は栄養や空間、温度、その他の条件による。その状態が変われば最適細胞数も変わる。例えば、栄養状態が悪くなれば最適細胞数も下がり、細胞の排出が起こる。細胞数が最適細胞数より少なかったり多かったりすると細胞は増えたり減ったりする。細胞数には変動がある。細胞数が最適数でない場合は最適数からの偏位である。現状の細胞数（N_C）は最適細胞数（N_p）からの偏位（N）と見なす。最適細胞数には散在性（σ）と揺動性（τ）が有るので存続性（H）や変動性（Z）もある。それゆえ実在性（Θ）がある。

　存在確率関数を適用する。

$$(72) \qquad \Phi = \exp[-(N_p - N_C)^2/(2\sigma^2)]/[\sigma(2\pi)^{1/2}]$$

σはΦの標準偏差値　σ > 0

偏位細胞数は $N = N_p - N_C$。

$$(73) \qquad \Phi = \exp[-N^2/(2\sigma^2)]/[\sigma(2\pi)^{1/2}]$$

細胞増殖率（R）は偏位数 N の微分である。

　持続確率分布は、

$$(74) \qquad \Psi = \exp\left[-R^2/(2\tau^2)\right]/\left[\tau(2\pi)^{1/2}\right]$$

⑵ 変動性

変動性（Z）は存在確率（Φ）と持続確率（Ψ）の比であるので、変動性の式（7）の対数を取り両側を微分する。

$$(11) \qquad (d\Phi/dt)/\Phi - (d\Psi/dt) - \Psi = (dZ/dt)/Z$$

式（73）と（74）を上式に代入すると、下式が得られる。

$$(75) \qquad (dN/dt)N/\sigma^2 - (dR/dt)R/\tau^2 = (dZ/dt)/Z$$

変動性（Z）は比例定数であり、増殖率 $R = dN/dt$ から下式を得る。

$$(76) \qquad d^2N/dt^2/\tau^2 - N/\sigma^2 = 0$$

$\tau/\sigma = \omega$ とすると、ω の次元は T^{-1} で角速度の次元である。

$$(77) \qquad d^2N/dt^2 = \omega^2 N$$

上式（77）を次式のように変換する。

$$(78) \qquad \left[d/dt + (\tau/\sigma)\right]\left[d/dt - (\tau/\sigma)\right]N = 0$$

右の項は拡散し現実的でないので、下式が適切な解となる。

$$(79) \qquad dN/dt + (\tau/\sigma)N = 0$$

この式の解は下式の減衰関数であり N_i は偏位細胞数 N の初期値である。

$$(80) \qquad N = N_i * \exp\left[-(\tau/\sigma)t\right]$$

式（80）は 0 に収束するから偏位細胞数は減衰因数（τ/σ）をもって減少し、0 に収束していく。

⑶ 存続性

存続性（H）は存在確率（Φ）持続確率（Ψ）の積であり、存続性（H）はその比例定数である。

$$(8) \qquad \Phi * \Psi = H$$

式（8）の対数を取り、両辺を微分して次式を得る。

$$(d\Phi/dt)/\Phi + (d\Psi/dt)/\Psi = dH/dt/H$$

式（73）と（74）を上式に代入する。次式を得る。

$$(dN/dt)N/\sigma^2 + (dR/dt)R/\tau^2 = dH/dt/H$$

N は N_p からの偏位、R = dN/dt そして H は定数、よって上式は次式となる。

$$(81) \qquad d^2N/dt^2/\tau^2 + N/\sigma^2 = 0$$

τ/σ は角速度（ω）であるから、上式は次式となる。

$$(82) \qquad d^2N/dt^2 = -\omega^2 N$$

この解は波動関数である。

⑷ 実在性

細胞は式（79）に示す変動性により増殖する。そして式（81）で示す存続性により安定化する。これら2式の和である実在性が増殖を示している。

変動性の式（79）を次のように変える。

$$(83) \qquad 2\omega dN/dt + 2\omega^2 N = 0$$

存続性も安定化に関与する。その式（82）を次の様式に変換する。

$$(84) \qquad d^2N/dt^2 + \omega^2 N = 0$$

偏位細胞数の実在性は式（83）と（84）の和であり、次式となる。

$$(85) \qquad d^2N/dt^2 + 2\omega dN/dt + 3\omega^2 N = 0$$

上式の解は次式となる。

$$N = \exp(-\omega t) * [N_1 * \cos(2^{1/2}\omega t) + \mathfrak{i}N_2 * \sin(2^{1/2}\omega t)] + C \quad C は定数$$

N は実数でなければならないから、$N_2 = 0$。

$t = \infty$ の場合は偏位細胞数 N は 0 だから $C = 0$。

$t = 0$ の時の N_1 は適正細胞数（N_p）と初期細胞数（N_i）の差であるから、$N_1 = N_p - N_i$、よって次式の減衰波動関数となる。現状細胞数 N_C は、$N_C = N_p - N$。

$$(86) \qquad N_C = N_p - (N_p - N_i) * \exp(-\omega t) \cos(2^{1/2}\omega t)$$

偏位細胞数 N を減少させていくと 0 になり、細胞数は（N_p）となり安定する。

⑸ オーバーシュート

　細胞培養では、しばしば細胞数が適正細胞数をオーバーシュートすることが観察される。細胞増殖が加速して適正細胞数を超えると細胞数が減少して適正細胞数以下になる。それは生化学上でのある未知の酵素による現象としばしば考えられていた。しかし、このオーバーシュートは安定性理論の減衰波動関数で説明できる。細胞は増殖する、その成長カーブは減衰関数カーブに似ている。ゴンペルツ関数モデルやロジスティック増殖モデルは式（80）が示す変動性による増殖カーブと似ている。これらの増殖モデルではオーバーシュートは説明されていない。細胞増殖において式（86）が示すような波動的増殖が、時には観察されて

いる。適正細胞数をオーバーシュートするのは細胞増殖の最初の波でしばしば観察されるが、その後の波は観察され難い。

2. 腫瘍増殖モデルと安定化原理

⑴ 腫瘍容積と活力

　生物学的分析では、しばしば細胞や組織の再生は内因的な誘因で起こると考えられている。成長率は組織の特性が変わらない限り一定であると考えられた。それが定率成長モデルである。それは簡単で短期間の腫瘍の成長を予測するのには便利である。細胞分裂や細胞消失は細胞に供給される栄養に大きく影響される。細胞死や中心壊死が増殖率の低下の原因となる。細胞には再生していく活力がある。しかし、この活力は細胞の内因的な誘因だけでなく栄養や他の刺激物等滋養物の取り込みにもよる。その環境から取り込む滋養物が腫瘍の活力の源である。滋養物は能動的な吸収や受動的に補充されている。同時に腫瘍が大きくなるにつれて細胞の排出や休眠が増加する。これにより滋養物の取り込みと排出量が同量となる。活力はこれらの増殖活動の総合であり、増加にも減少にもなり得る。活力は腫瘍が大きくなるほど下がると考えられていたから、一般に活力は腫瘍体積（G）に反比例すると考えられていた。そして同時に、活力は増殖活動を高めているから活力は増殖率に比例する。よって増殖活力（P）とは比増殖率と名付け得る。腫瘍容積はこの増殖活力の蓄積と考えられる。増殖活力の積分を生育力（Q）と名付ける。生育力が腫瘍容積を増加させる。それは腫瘍容積（G）の対数に比例している。腫瘍の増殖性は生育力による。

⑵ 腫瘍容積と生育力

　生育力が腫瘍体積を変化させるということは腫瘍体積がその生育力の従属変数であり変動性を応用することにより明らかとなる。dG/dQ は

生育力による腫瘍の増殖性 (G_q) とする。腫瘍容積 (G) には散在性 (σ) があり存在確率分布 (Φ) を持つ。また増殖性 (G_q) には揺動性 (τ) があり持続確率分布 (Ψ) を持つ。それらの比は変動性 (Z) を成す。

$$(87) \qquad \Phi = \exp[-G^2/(2\sigma^2)]/[\sigma(2\pi)^{1/2}]$$

σはΦの標準偏差　σ > 0

腫瘍の増殖性 (G_q) の持続確率分布は、

$$(88) \qquad \Psi = \exp[-G_q^2/(2\tau^2)]/[\tau(2\pi)^{1/2}]$$

τはΨの標準偏差　τ > 0

変動性 (Z) は存在確率分布 (Φ) と持続確率分布 (Ψ) の比であるから、

$$Z = \Phi/\Psi$$

上式の両辺の対数を取り、それを Q に関して微分する。

$$(d\Phi/dQ)/\Phi - (d\Psi/dQ)/\Psi = (dZ/dQ)/Z$$

上式に式 (87) と (88) を代入すると、次式を得る。

$$(89) \qquad (dG/dQ)G/\sigma^2 - (dG_q/dQ)G_q/\tau^2 = (dZ/dQ)/Z$$

変動性 (Z) は比例定数であり、$G_q = dG/dQ$ であるから次式を得る。

$$(90) \qquad d^2G/dQ^2/\tau^2 - G/\sigma^2 = 0$$

上式 (90) は次式の如く変更できる。

$$[d/dQ+(\tau/\sigma)][d/dQ-(\tau/\sigma)]G=0$$

$dG/dt+(\tau/\sigma)G$ は 0 に収束し現実性はないので、次式を採用する。

(91) $dG/dQ-(\tau/\sigma)G=0$

上式の解が次式であるから、その対数を取れば式（92）を得る。

$$G=\exp[(\tau/\sigma)Q]$$

(92) $\mathrm{Log}(G)=(\tau/\sigma)Q$

生育力は腫瘍容積の対数に比例し、散在性や揺動性に関与する比例定数が付く。

　上式を時間にて微分する。増殖活力（P）は dQ/dt であるから次式を得る。

$$P=dG/dt/G/(\tau/\sigma)$$

増殖活力は上述の如く増殖率に比例して、腫瘍容積に反比例する。

　生育力には内因的誘因とその環境のバランスに基づいた適正値（Q_p）がある。

　次式が生育力の適正値（Q_p）からの偏位（Q_d）を示す。

$$Q_d=Q-Q_p$$

生育力には散在性（σ）の存在確率分布がある。増殖活力（P）には揺動性（τ）の持続確率分布がある。よって、存続性（H）や変動性（Z）、実在性（Θ）がある。

(93) $\Phi=\exp[-(Q-Q_p)^2/(2\sigma^2)]/[\sigma(2\pi)^{1/2}]$

σはΦの標準偏差 σ > 0

（Q–Q_p）は偏位（Q_d）であるから、

$$(94) \qquad \Phi = \exp\left[-Q_d^2/(2\sigma^2)\right]/\left[\sigma(2\pi)^{1/2}\right]$$

増殖活力（P）は生育力（Q）の微分であるから、次式の持続確率分布がある。

$$(95) \qquad \Psi = \exp\left[-P^2/(2\tau^2)\right]/\left[\tau(2\pi)^{1/2}\right]$$

変動性（Z）は存在確率分布（Φ）と持続確率分布（Ψ）の比であり、次式で示される。

$$Z = \Phi/\Psi$$

両辺の対数を取り、微分すると次式となる。

$$(d\Phi/dt)/\Phi - (d\Psi/dt)/\Psi = (dZ/dt)/Z$$

式（94）と（95）を上式に代入すると、次式を得る。

$$(dQ_d/dt)Q_d/\sigma^2 - (dP/dt)P/\tau^2 = (dZ/dt)/Z$$

変動性（Z）は比例定数であり、P = dQ/dt であるから次式を得る。

$$(96) \qquad d^2Q_d/dt^2/\tau^2 - Q_d/\sigma^2 = 0$$

上式を次式のように変更する。

$$\left[d/dt + (\tau/\sigma)\right]\left[d/dt - (\tau/\sigma)\right]Q_d = 0$$

$dQ_d/dt - (\tau/\sigma Q_d)$ は拡散するので現実性がない、現実性のある解は次式である。

$$(97) \qquad dQ_d/dt+(\tau/\sigma)\,Q_d = 0$$

この式の解は下記の減衰関数であり、Q_i は偏位 Q_d の初期値である。τ/σ は角速度（ω）に置き換えられる。

$$(98) \qquad Q_d = Q_i*\exp(-\omega t)$$

偏位は減衰係数（$\omega = \tau/\sigma$）にて減少し、0 に収束し生育力（Q）は適正値 Q_p となる。

　存続性（H）は存在確率（Φ）と持続確率（Ψ）の積である。存続性（H）はその比例定数である。その式は、

$$\Phi*\Psi = H$$

上式の対数を取り、微分する。

$$(15) \qquad (d\Phi/dt)/\Phi + (d\Psi/dt)/\Psi = dH/dt/H$$

式（94）と（95）を上式に代入し、次式を得る。

$$(dQ_d/dt)\,Q_d/\sigma^2+(dP/dt)\,P/\tau^2 = dH/dt/H$$

Q_d は Q_p からの偏位で、$P = dQ_d/dt$、そして H は定数。よって、上式は次式となる。

$$d^2Q_d/dt^2/\tau^2+Q_d/\sigma^2 = 0$$

この式は振動関数で、τ/σ は角速度（ω）であり、次式の波動関数となる。

$$(99) \qquad d^2Q_d/dt^2 = -\omega^2Q_d$$

細胞は式（97）が示す如く変動性に基づき増加し、この式（99）が示す

如く存続性に基づき維持される。これら二式の和による実在性に基づい
て増殖している。

式（97）は変動性の式で、次のように変える。

$$(100) \qquad 2\omega dQ_d/dt + 2\omega^2 Q_d = 0$$

次式の存続性は安定に寄与する。

$$(99) \qquad d^2 Q_d/dt^2 + \omega^2 Q_d = 0$$

式（10）による増殖の実在性は式（100）と（99）の和であり、その結
果は次式である。

$$(101) \qquad d^2 Q_d/dt^2 + 2\omega dQ_d/dt + 3\omega^2 Q_d = 0$$

この式の解は次式である。

$$Q_d = \exp(-\omega t) * [Q_1 * \cos(2^{1/2}\omega t) + i Q_2 * \sin(2^{1/2}\omega t)]$$

Q_d は実数であるから $Q_2 = 0$。$t = 0$ の時、$Q_1 = Q_{io}$。Q_i は初期偏位、する
と、

$$Q_d = Q_i * \exp(-\omega t) * \cos(2^{1/2}\omega t)$$

内因的誘因や外因的条件により種々な適正生存力が起こり得る。実際に
適正生育力（Q_p）を求めるには最小二乗法等によらなければならない。
減衰波動関数式（102）が応用し易いであろう。

$$(102) \qquad Q = Q_p - Q_i \exp(-\omega t)\cos(2^{1/2}\omega t)$$

波高 Q_i は減少して 0 に近づき増殖は安定する。腫瘍体積で表すなら生
育力は体積の対数に比例しているので次式となる。

$$(103) \qquad \log(G) = \log(G_p) - \log(G_i)\exp(-\omega t)\cos(2^{1/2}\omega t)$$

G_p は適正腫瘍体積、G_i は初期腫瘍体積で表した初期波高（初期偏位）。

３．人間の寿命とゴンペルツ関数

⑴ ゴンペルツ関数

　1825年、ベンジャミン・ゴンペルツは人間の死亡率に関する数式モデルを発表した。彼が人間の年齢と死亡率に関して見つけたことは、人の年齢の指数関数と死亡率の対数関数が直線関係となるということである。分析結果として、次のように述べている。

「人間は平均的に死を逃れるための力を均等な微小時間内に均等割合で消耗して破滅に立ち向かっているのであろう。」

　ゴンペルツ関数は一般に、次の式で示される。

$$（104）\qquad Y = AB^{\wedge}c^{\wedge}x$$

Y は死亡率、A は基礎死亡率で定数、c は定数、x は年齢、B は年齢依存定数で０歳児の値。ゴンペルツ関数にて仮定されている初期死亡率はA*B。

　見易いように上式を対数表示する。

$$（105）\qquad \log Y = \log A + \log B * \exp（x * \log c）$$

上式に示されているように、死亡率（Y）の対数と年齢（x）の指数は直線関係である。死亡率は年齢とともに上昇する。上述の「破滅に立ち向かう力」とは生きる活力を意味する。事実、活力の喪失は衰弱を進行させる。衰弱は活力の低下により発生するゆえ、活力の従属変数である。年齢は活力の低下を起こし衰弱を進行させる。衰弱と活力は年齢の従属変数である。

⑵ 死亡率と衰弱

　死亡率は衰弱により変化する。だから衰弱は死亡率の存在確率と持続確率の基となる変数である。衰弱（y）による死亡率の存在確率と持続確率を安定性理論の変動性を応用することにより、衰弱から死亡率を得られる。死亡率の存在には存在確率と散在性がある。Y を死亡率とする。

$$(106) \qquad \Phi = \exp[-Y^2/(2\sigma^2)]/[\sigma(2\pi)^{1/2}]$$

死亡率には変化があり、持続確率分布と揺動性を伴う。F を衰弱による死亡率変化とする。

$$(107) \qquad \Psi = \exp[-F^2/(2\tau^2)]/[\tau(2\pi)^{1/2}]$$

変動性は存在確率（Φ）と持続確率（Ψ）の比であり、その式の対数を取り微分したのが次式である。

$$(11) \qquad (d\Phi/dy)/\Phi - (d\Psi/dy)/\Psi = (dZ/dy)/Z$$

式（106）と（107）を上式に代入すると、次式を得る。

$$(108) \qquad (dY/dy)Y/\sigma^2 - (dF/dy)F/\tau^2 = (dZ/dy)/Z$$

変動性（Z）は比例定数であり、死亡率変動は $F = dY/dy$。よって、次式を得る。

$$(109) \qquad d^2Y/dy^2/\tau^2 - Y/\sigma^2 = 0$$

上式（109）は次のように変え得る。

$$[d/dy + (\tau/\sigma)][d/dy - (\tau/\sigma)]Y = 0$$

$[dY/dy + (\tau/\sigma)Y = 0]$ の Y は減少し、0 に収束する。それは死亡率にはあり得ない。衰弱により上昇する過程であるから、右項の式が適用される。

$$(110) \qquad dY/dy - (\tau/\sigma)Y = 0$$

この式の解は次式である。

$$(111) \qquad Y = Y_0 * \exp[(\tau/\sigma)y]$$

$$(112) \qquad \log Y = \log Y_0 + (\tau/\sigma)y$$

⑶ 衰弱の変動性

衰弱は年齢による。年齢が衰弱に大きな影響を与える。よって、年齢（x）が衰弱の存在と持続を引き起こしている。

衰弱の数式は安定性理論の変動性を応用することにより得られる。衰弱には存在確率分布と持続確率分布がある。y を年齢による衰弱とする。

$$(113) \qquad \Phi = \exp[-y^2/(2\sigma^2)]/[\sigma(2\pi)^{1/2}]$$

活力の喪失が衰弱を変化させる。その活力（V）には持続確率分布と揺動性がある。

$$(114) \qquad \Psi = \exp[-V^2/(2\tau^2)]/[\tau(2\pi)^{1/2}]$$

存在確率（Φ）と持続確率（Ψ）比である変動性を求め、その式の対数を取り微分すると次式を得る。

$$(11) \qquad (d\Phi/dx)/\Phi - (d\Psi/dx)/\Psi = (dZ/dx)/Z \qquad x \text{は年齢}$$

式（113）と（114）を上式に代入すると、次式を得る。

$$(115) \qquad (dy/dx)y/\sigma^2 - (dV/dx)V/\tau^2 = (dZ/dt)/Z$$

変動性（Z）は比例定数であり、活力は $V = -dy/dx$、そして $dV/dx = -d^2y/dx^2$。よって次式を得る。

(116)　　　$d^2y/dx^2/\tau^2 - y/\sigma^2 = 0$

上式（116）を次のように変形する。

$$[d/dx + (\tau/\sigma)][d/dx - (\tau/\sigma)]y = 0$$

左項［$dy/dx + (\tau/\sigma)y = 0$］の衰弱 y は減少して 0 に収束するので現実に
はない。これは悪くなる過程なので右項の次式が解である。

(117)　　　$dy/dx - (\tau/\sigma)y = 0$

この式の解は次式である。

$$y = y_0 * \exp[(\tau/\sigma)x]$$

式（112）に、この式を代入する。

(112)　　　$\log Y = \log Y_0 + (\tau/\sigma)y$

(118)　　　$\log Y = \log Y_0 + (\tau/\sigma) * y_0 * \exp[(\tau/\sigma)x]$

$\log Y_0$ は定数で $\log A$ に当たり、$(\tau/\sigma) * y_0$ は衰弱 y の初期値で $\log B$ に当
たり、τ/σ は $\log c$ だから、

$$Y = A * B ^\wedge \exp(x * \log c)$$

ゴンペルツ関数では人の寿命の存続性は無視されている。存続性を考慮
すれば死亡率曲線波動性を含んで上昇するであろう。

4．ゴンペルツ関数腫瘍成長モデル

⑴ ゴンペルツ関数の応用

　ゴンペルツ関数は安定性理論の変動性に基づいているので色々な現象

に適用できる。それは1825年に人間の死亡率を推定する為に開発された。1926年にライトとスォールが『アメリカ統計学会誌』の概説でゴンペルツ関数は細胞増殖や臓器の成長等生物学的な成長に適用できることを発表している。死亡曲線では指数の項が陽性であるが、生物学的な成長曲線ではその指数の項が陰性となる。腫瘍増殖モデルに下記の対数表示のゴンペルツ関数を応用する。

$$\log(Y) = \log(A) + \exp(x*\log c)\log(B)$$

$\log(Y)$ を腫瘍体積 G に、x を時間 t に、$\log c$ は陰性定数であるから $(-r)$ に置き換える。

$t = \infty$ の時、腫瘍体積 G は最大、即ち適正体積 (G_p) だから、$\log(A) = G_{po}$ 初期 $t = 0$ で腫瘍体積 G は初期体積 (G_i) だから、$\log(B) = -(G_p - G_i)$。

適正体積 G_p からの偏位体積を D とし初期偏位体積を D_i とすると、$G_p - G_i = D_{io}$。よって、上式は次のように書き換えられる。

(119)　　　$G = G_p - D_i*\exp(-r*t)$

腫瘍体積は減衰係数 $(-r)$ にて偏位体積が減少し適正体積 G_p に収束する。

腫瘍体積 G に次節のように変動性を適用すると同じ等式が得られる。

ゴンペルツ関数は消滅への変動性を表す関数であるが腫瘍の成長は増加の変動性を表す関数として用いるのであるから変化を表す項の係数は逆になる。

⑵ ゴンペルツ関数における変動性

腫瘍体積の変動を求めるのであるが、上述のように定数である適正腫瘍体積と偏位体積の差が腫瘍体積である。腫瘍体積の変動は偏位体積の変動と同一であるゆえ偏位体積の変動を検討する。偏位体積の存在には

存在確率分布とその散在性（σ）を伴う。

$$(120) \qquad \Phi = \exp[-D^2/(2\sigma^2)]/[\sigma(2\pi)^{1/2}]$$

そして増殖には持続確率分布とその揺動性（τ）を伴う。R は偏位変動率。

$$(121) \qquad \Psi = \exp[-R^2/(2\tau^2)]/[\tau(2\pi)^{1/2}]$$

変動性（Z）を適用する。それは存在確率（Φ）と持続確率（Ψ）の比である。

$$\Phi/\Psi = Z$$

この式の対数を取り、それを微分したのが次式である。

$$(11) \qquad (d\Phi/dt)/\Phi - (d\Psi/dt)/\Psi = (dZ/dt)/Z$$

これに式（120）と（121）を代入したのが次式である。

$$(122) \qquad (dD/dt)D/\sigma^2 - (dR/dt)R/\tau^2 = (dZ/dt)/Z$$

変動性（Z）は比例定数。偏位変動 R は R = dD/dt である。
よって、次式を得る。

$$(123) \qquad d^2D/dt^2/\tau^2 - D/\sigma^2 = 0$$

上式は次のように変形できる。

$$[d/dt+(\tau/\sigma)][d/dt-(\tau/\sigma)]D = 0$$

[dD/dt−(τ/σ)D = 0] は拡散する式であるから現実的ではない。よって次式が解となる。

$$(124) \qquad dD/dt+(\tau/\sigma)D = 0$$

$$D = D_i * \exp\left[(-\tau/\sigma)t\right]$$

D_i は初期偏位（G_p-G_i）で、G_i は初期腫瘍体積。

　腫瘍体積 G は増加関数であり最大値 G_p（適正体積）からの偏位 D として、

$$(125) \qquad G = G_p - D_i * \exp\left[(-\tau/\sigma)t\right]$$

ゴンペルツ関数を用いた腫瘍増殖式と同じ内容である。（τ/σ）は増殖率 r（$r=-R$）。収束していく変動性では指数関数の係数も指数も負の定数である。寿命の如く破壊的で拡散する変動性ではいずれの定数も正となる。

⑶ 腫瘍増殖の存続性

　腫瘍の増殖は存続性にも影響される。存続性（H）は存在確率（Φ）と持続確率（Ψ）の積であり、存続性（H）が比例定数である。

$$(8) \qquad \Phi * \Psi = H$$

この式の対数を取り、微分した式が次式である。

$$(d\Phi/dt)/\Phi + (d\Psi/dt)/\Psi = dH/dt/H$$

式（120）と（121）を上式に代入すると次式を得る。

$$(126) \qquad (dD/dt)D/\sigma^2 + (dR/dt)R/\tau^2 = dH/dt/H$$

$R = dD/dt$ であり H は比例定数。よって、上式は次式となる。

$$(127) \qquad d^2D/dt^2/\tau^2 + D/\sigma^2 = 0$$

散在性と揺動性の比（τ/σ）は角速度（ω）であり、上式は次式の波動関数である。

(128)　　　$d^2D/dt^2 = -\omega^2 D$

⑷ 腫瘍増殖の実在性

　実在性は変動性と存続性の積であり、式（124）と（128）の和である。

　変動性の式（124）を次式のように2倍にする。

(129)　　　$2\omega dD/dt + 2\omega^2 D = 0$

存続性の式（128）も安定化に関与する。

(128)　　　$d^2D/dt^2 + \omega^2 D = 0$

増殖の実在性は式（10）により式（129）と（128）の和であるから、次式となる。

(130)　　　$d^2D/dt^2 + 2\omega dD/dt + 3\omega^2 D = 0$

上式の解は次のようになる。

$$D = \exp(-\omega t) * [C_1 * \cos(2^{1/2}\omega t) + iC_2 * \sin(2^{1/2}\omega t)] + C$$

C_1, C_2, Cは定数

D は実数であるから $C_2 = 0$、$t = 0$の時、$C_1 = D_i$（初期偏位）であるから、

$$D = D_i * \exp(-\omega t) * \cos(2^{1/2}\omega t)$$

腫瘍体積で表すなら $D_i = G_p - G_i$ であり、$G = G_p - D$ であるから、次式の減衰波動関数となる。

(131)　　　$G = G_p - (G_p - G_i)\exp(-\omega t)\cos(2^{1/2}\omega t)$

振幅 D は減少し 0 に近づくと腫瘍増殖は安定して適正体積 G_p となる。現実に G_p を求めるには最小二乗法を用いなくてはならない。その場合、安定化過程の波動性や腫瘍増殖のオーバーシュートが無視される可能性がある。

エピローグ

そもそも古くからの人間の考える概念は、神の摂理により自然が、またはその大部分が創造されたとする。命題として神の摂理である自然は完全で不安定性はなく絶対であると信仰された。しかし、自然は人々の知力による研究でこの命題に反する各種法則等が見出されている。だが未だ包括的な法則は出ていない。自然はその全部、またはその大部分が偶然により創造されてきたことは反定立となるが、偶然性のもたらす法則に自然は基づいている。その偶然性こそ神の摂理であると考えるならこの命題は成立する。この法則の原理は安定化であり、安定化原理を用いて基本法則を説明することができる。筆者は医学を目指して大学に入学したのであるが、図書室で最初に手にした本でロレンツ関数を見た。この不思議な現象を信じるのは苦痛であった。長い年月の末、偶然の現象が安定化するのが答えと考えた。この安定性をもたらす原理は正規関数で表す偶然性にあった。筆者が腫瘍増殖の数式モデルを求め研究データを解析している時に、成長カーブに波動関数が含まれているのではないかと気付いたが、議論の場では常に測定誤差として処理されていた。しかし、細胞培養の初期にオーバーシュートするのは波動関数の関与で説明できるので、長年波動関数の関与を信じて追求し、安定性を正規関数で表すと説明できることが分かった。存在状況が変化すると、物は変化して安定化する。存在し易い安定な状態に変動する。この原理で多くの事象を説明できる。腫瘍の増殖や死亡などの生物事象もこの安定化原理で説明できることを示した。原理的な引力の存在を説明できることが分かり、その後は初等力学に応用して追求を続けていたのである。引力の成因は散在性にある。特殊相対性の成因は揺動性にある。この二つの素因が変動性と存続性をもたらす。この二つの成因の調和に基づき実在性をもたらしている。この実在性は質量が有る事を前提としない。空間

は実在するゆえ実在性が有る。空間内のエネルギーも実在する。質量を持たない中間子等も実在している。

　単純には物の安定性について次のことが言える。

　　「壊れ易い物は存在し難いが、適切な条件下では存在する。」

　存続性の低い事象は変動性が高く変化するが、存在状況が変化し変動性の低い状況を得れば存続性は高まるということである。変動性と存続性を分析することにより上記の二つの原理は導かれたのである。事象の実在は変動性と存続性が反比例して存在の実在性をもたらしていると認識するべきである。

　この偶然性に基づく安定化原理は各種基本法則を説明することができた。これらの問題に答えてくれたこの原理は多くの事象に応用可能であり、諸種の研究に利用されたい。

赤沼　篤夫（あかぬま　あつお）

1964年３月東京大学医学部医学科を卒業後、放射線医学教室に入り放射線の研究を始める。1968～1973年まで米国とドイツに留学し、エッセン大学助教やストーニイブルック大学医学部講師を兼務。ブルックヘブン国立研究所で客員研究員を併任し、粒子線の研究を行う。1973年帰国し東京大学講師。1978年には米国ロスアラモス国立研究所でパイオンの医学利用の研究に携わる。1985年東京大学準教授に。重粒子線の研究のため1991年放射線医学総合研究所に転勤したが研究が進まず、独自に研究するため1997年退官。現在に至る。

安定化原理
Stabilization Principle

2020年10月31日　初版第１刷発行

著　　者　赤沼篤夫
発行者　中田典昭
発行所　東京図書出版
発行発売　株式会社 リフレ出版
　　　　　〒113-0021　東京都文京区本駒込 3-10-4
　　　　　電話 (03)3823-9171　FAX 0120-41-8080
印　　刷　株式会社 ブレイン

© Atsuo Akanuma
ISBN978-4-86641-348-8 C3042
Printed in Japan 2020

落丁・乱丁はお取替えいたします。
ご意見、ご感想をお寄せ下さい。